AuthorHouse™ LLC
1663 Liberty Drive
Bloomington, IN 47403
www.authorhouse.com
Phone: 1-800-839-8640

© 2014 Francis Mensah. All rights reserved.

No part of this book may be reproduced, stored in a retrieval system, or transmitted by any means without the written permission of the author.

Published by AuthorHouse   05/13/2014

ISBN: 978-1-4969-0584-0 (sc)
ISBN: 978-1-4969-0583-3 (hc)
ISBN: 978-1-4969-0719-6 (e)

Library of Congress Control Number: 2014907258

Any people depicted in stock imagery provided by Thinkstock are models, and such images are being used for illustrative purposes only. Certain stock imagery © Thinkstock.

This book is printed on acid-free paper.

Because of the dynamic nature of the Internet, any web addresses or links contained in this book may have changed since publication and may no longer be valid. The views expressed in this work are solely those of the author and do not necessarily reflect the views of the publisher, and the publisher hereby disclaims any responsibility for them.

Topics in Optics and Laser Light in the Atmosphere

By Francis E. Mensah

Preface

This book covers topics in optics and the use of laser light in the atmosphere. I expect that, at least, it would benefit number of undergraduate and graduate students in the field of optics, lidar, laser instrumentation, and data analysis. It would also benefit ordinary people who are interesting in optics and remote sensing with laser. The first chapter starts with a review of Newton's laws and planetary motion. The second chapter deals with the Planet earth's atmosphere, the third is an introduction to the remote sensing. Chapter 4 and 5 introduce a background on Maxwell's laws in electromagnetism and light polarization. Some other topics of interest have been also developed. Among these topics are the light interaction with spherical surfaces and related equations, light Interference, linear polarization by anisotropy, Fourier transform spectroscopy, and an introduction to lidar. However, readers can read this book differently and draw their own conclusions. I welcome suggestions and discussions. I don't pretend to have covered all subjects related to the topic. But the very basic concept presented in this book is intended to help everyone who would like to do research involving

optics and lidar. I also would like to acknowledge that like many books, this book does not come from a vacuum and I am indebted to many scientists who have done some research in the past on the fields of optics and laser light.
Francis E.T. Mensah, PhD.

Acknowledgement

I would first like to express my sincere gratitude to the Department of Physics and Astronomy at Howard University. It is my great pleasure to acknowledge the help and cooperation I have received from Professor Arthur Thorpe in the Department of Physics and Astronomy at Howard University. I am also very grateful to Professor Demetrius Venable of the department of Physics and Astronomy at Howard University and at the Howard University Beltsville campus for his important contributions. I express my gratitude to Dr. Franck Senftle and to Dr. Marcus Alfred. My thanks also go to Mr. Julius Grant for his constant services with the software Easy-Plot^{TM}.

There are many other people that I will not be able to list here but with whom I have had useful and inspiring conversations about this book.

Finally I am greatly indebted to my wife Edwige for her supports.

Francis E. Mensah

Chapter 1
Review of Newton's laws
1.1. Introduction
1.2. Newton's first law of motion
1.3. Newton's second law of motion
1.4. Newton's third law
1.5. Newton's law of universal gravitation
1.6. Variation of the gravitation acceleration with the altitude
1.7. Planetary motion and Kepler's laws

Chapter 2 The Planet earth's atmosphere
2.1. Description of the solar system
2.2. Planet Earth atmosphere composition.
2.3. Particles in the troposphere
2.3.1. Sea Salt particles
2.3.1. Sea Salt particles
2.3.2. Particles from fires, rocks, soils and volcanoes

Chapter 3
Remote sensing
3.1. Introduction
3.2. History of remote sensing
3.3. Passive and active systems

3.4. Passive sensing
3.4.1. Beer's law and remote sensing of aerosol
3.4.1.1. Lambert's law

Chapter 4
Background on Maxwell's laws in electromagnetism
4.1. Maxwell equations
4.2. Equation of conservation of charge
4.3. Electromagnetic energy
4.3.1. Electrostatics energy
4.3.2. Magnetic energy
4.3.3. The Poynting vector
4.3.4. The total energy in a linear medium
4.4. Energy conservation
4.5. The Electromagnetic wave equations
4.6. Electromagnetism wave equations in an inhomogeneous medium
4.7. Poisson and Laplace equations

Chapter 5
Light polarization in a given medium
5.1. State of polarization
5.2. The linear polarization
5.3. Circular polarization

5.4. Elliptical polarization

5.5. Natural and partially polarized light

5.6. Matrix representation of polarization

5.7. Jones vector for linearly polarized light

5.8. Jones vector for circular polarization

5.9. Jones matrix for a rotator

Chapter 6
Review of Geometric optics

6.1. Introduction

6.2. The thin lens model

6.3. Laws of geometrical optics

6.3.1. Law of Reflection

6.3.2. The Law refraction

6.4. The law of refraction and Fermat's principle

6.5. Light interaction with a prism

Chapter 7
Light interaction with spherical surfaces and related equations

7.1. Light and the convex spherical surfaces

7.2. Object and Image distance: Analysis at the infinity

7.2.1. Object at the infinity

7.2.2. Object between the focal length image F' and the optical center O
7.2.3. Magnification
7.3. Concave spherical surface
7.3.1. Thin lens equation
7.3.2. Spherical mirror

Chapter 8
Light Interference
8.1. Introduction
8.2. Harmonic waves
8.3 Superposition of the harmonic waves
8.4. Interferences intensities
8.5. Young experiment
8.6. Newton's rings
8.7. The Michelson interferometer
8.8. Fresnel's formulas
8.9. Blackbody radiation

Chapter 9
Linear polarization by anisotropy
9.1. Introduction
9.2. Birefringence

Chapter 10

Fourier transform spectroscopy

10.1. Introduction

10.2. Fourier series

10.3. Fourier transform

10.3.1. Definition of Fourier transform

10.3.2. Properties of the Fourier transform

10.4. Fourier transform spectroscopy

Chapter 11

Laser optics

11.1. Introduction and history of laser

11.2. Theory and definition of laser

11.3. Spontaneous emission

11.4. Stimulated emission

11.5. Absorption

11.6. Laser idea and population inversion

11.7. Pumping schemes

11.8. Different types of lasers

11.8.1 Solid state lasers

11.8.1.1. The ruby laser

11.8.1.2. The neodymium laser

11.8.2. Gas lasers and excimers

11.8.2.1. Helium-Neon laser

11.8.2.2. Excimer laser

11.9 Light scattering by molecules: Raleigh scattering

11.10. Light scattering by particulates: Mie scattering

11.11. Inelastic scattering: Raman scattering

Chapter 12
Lidar theory
12.1. Introduction and definition of lidar
12.2. The lidar equation
12.3. Some basic definitions relative to water vapor
12.3.1. Water vapor mixing ratio
12.3.2. Relative humidity
12.3.3. Specific humidity
12.3.4. Water vapor density
12.3.5. Absolute humidity
12.3.6. Partial pressure and vapor pressure
12.3.7. The wet-bulb temperature
12.4. Overview of lidar techniques used by scientists

Chapter 13
Equipment and experimental procedure
13.1. Introduction
13.1.1. The transmitter system

13.1.1.1. The example of the 248 nm KrF laser system

13.1.1.2. Characteristics of the 355 nm Nd:YAG lidar system

13.1.2. The receiver

13.1.3. The detector system

13.1.3.1. The photomultiplier tube, PMT

13.1.3.2. The spectrometer

13.1.3.3. The beam acquisition electronics

13.1.3.4. The filter

13.2. Experimental procedure

13.2.1. The block diagram of the lidar system

13.2.2. The SOLEX method

Chapter 14

Calibration and introduction to data analysis

14.1. Introduction to Calibration and data analysis

14.2. Data signal from high resolution Tektronix 7104 oscilloscope.

Appendix

Appendix A:
Appendix B
References

Chapter 1

Review of Newton's laws

1.1. Introduction

1.2. Newton's first law of motion

1.3. Newton's second law of motion

1.4. Newton's third law

1.5. Newton's law of universal gravitation

1.6. Variation of the gravitation acceleration with the altitude

1.7. Planetary motion and Kepler's laws

1.1. Introduction
The review of Newton's law is important at the beginning of this book because they deal with motion and gravitation.

1.2. Newton's first law of motion
Newton's first law of motion also called the law of inertia describes the tendency of an object to maintain it original states of motion called inertia. The law is stated as follows: An object at rest stays at rest as far as the net force acting upon it is zero. Otherwise, if it moves, it will move with a constant velocity unless acting upon by an unbalanced net force. In this case the acceleration of the center of mass is zero:

$$\vec{a} = \vec{0} \qquad (1.1)$$

1.3. Newton's second law of motion
Newton second law of motion concerns the force and the acceleration of an object in motion. According to this law, and object in which is acted upon a non-zero net force moves with an acceleration that is proportional to the net force and

inversely proportional to the mass of the object. The acceleration in this case has the same direction as the net force. Newton's second law motion can be summarize in the following equation:

$$\vec{a} = \frac{\vec{F}}{m} \qquad (1.2)$$

Newton second law of motion can also be written in term of the change in the linear momentum, $\vec{P}$:

$$\vec{F} = \frac{d\vec{P}}{dt} \qquad (1.3)$$

Note that the linear momentum of a particle or object of mass m moving with a velocity $\vec{v}$ is given by $\vec{P} = m\vec{v}$

1.4. Newton's third law

Newton's third law of motion is the law of action and reaction. According to this law, for any action, there is an equal and opposite reaction. So, if we design the action and reaction respectively by $\vec{A}$ and $\vec{R}$, we will have the relation:

$$\vec{R} = -\vec{A} \qquad (1.4)$$

1.5. Newton's law of universal gravitation

Newton developed this theory in 1965 for some reason it would not be publish until 1685. The most important aspect of this law was that the gravitational force is universal and that the same force that caused apples to fall was also responsible for the motion of the moon. The law states that any two bodies in the universe attract each other by a common force that is proportional to the masses of the bodies and inversely proportional to the distance separating the two bodies. The coefficient of proportionality is called the universal constant of gravitation. Mathematically the law can be given by the equation:

$$F = G\frac{mM}{r^2} \qquad (1.5)$$

The constant of proportionality G is given by:
$$G = 6.673 \times 10^{-11} Nm^2/kg^2 \qquad (1.6)$$

Figure: 1.1. Newton's law of universal gravitation between two bodies

1.6. Variation of the gravitation acceleration with the altitude

Consider the body of mass M as the planet Earth and the object of mass m as an apple. The apple is at a distance r from the center of mass of the planet Earth. We assume the planet Earth to be spherical (see figure 1.2).

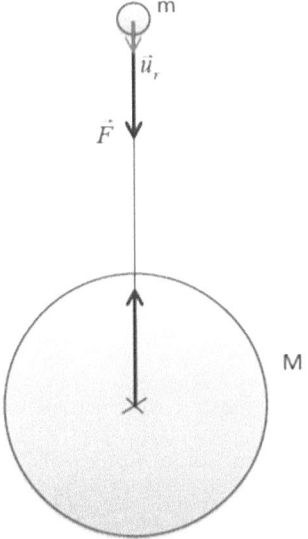

Figure: 1.2. Interaction between an object of mass m and the planet Earth.

Now, let us use the two formulas related to Newton's law of universal gravitation and Newton's second law of motion:

$$\vec{F} = G\frac{mM}{r^2}\vec{u}_r \tag{1.7}$$

$$\vec{F} = m\vec{a} \tag{1.8}$$

$\vec{u}_r$ is a unit vector that has the same direction as the one of the gravitation force which is directed to the center of the planet Earth. Those two forces are equal. Then we have:

$$\vec{a} = G\frac{M}{r^2}\vec{u}_r \qquad (1.9)$$

In magnitude,

$$a = G\frac{M}{r^2}. \qquad (1.10)$$

If the radius of the planet Earth is R_e, the acceleration on the surface of the planet Earth is $g = G\dfrac{M}{R_e^2}$. Comparing the acceleration on the surface of the planet Earth and the one at the distance r from the center of the planet Earth, we have:

$$\frac{a}{g} = \left(\frac{R_e}{r}\right)^2 \qquad (1.11)$$

1.7. Planetary motion and Kepler's laws

For centuries, physicists have been trying to explain the motion of the planets around the sun in the solar system. Among the theories that held sway over years, was the geocentric theory associated with Claudius Ptolemy (C. A.D. 150). This theory was successful in explaining the planetary motion with a degree of accuracy related to the level of knowledge in those days. In this theory, the planet Earth was the center of the universe. In fact, the word geocentric can be split in two words: geo which means Earth and centric or center. This theory gives a model called the Ptolemaic model and was accepted by old civilizations such as the Greek civilization. It is then assumed that all the celestial bodies such as the sun, the moon, the stars and others, circled around the Earth.

Before Ptolemy, other geek philosophers, Aristotle and his student Plato wrote on the theory of geocentrism considering the planet Earth as spherical, stationary at the center of the universe and other celestial bodies turn around the earth on circular orbits. A century later, Aristarchus of Samos (310 -210 B.C.), propose his theory considering the sun fixed at the center of the universe. In his theory the earth revolve around the sun in a circular orbit. He also noticed that the start appears fixed in position because their distances from the sun were huge compare with the distance sun-earth. This theory called heliocentric theory was not well accepted in those days. From the second century to the sixteen century only the Ptolemy theory was accepted and taught. Later, Nicolas Copernicus (1472-1543) will revive again the heliocentric theory of Aristarchus of Samos which will trigger a scientific revolution in which other great physicist such as Kepler, Galileo and Newton would play a key role. In his heliocentric theory, Copernicus considers the sun at the center of the universe and all the planets revolve around the sun in circular orbits. The fixed stars were assumed to lie in spheres surrounding the solar system. The heliocentric theory met some opposition. However, a famous Danish astronomer, Tycho Brahe (1546-1601) made very careful and unprecedented accurate measurements of the motions of the planets that convinced him about Copernican hypothesis. Brahe careful and convincible observations and measurement were bequeathed to German astronomer Johannes Kepler (1571-1630) who after analysis of the data introduced three laws that describe the motions of the planets around the sun.

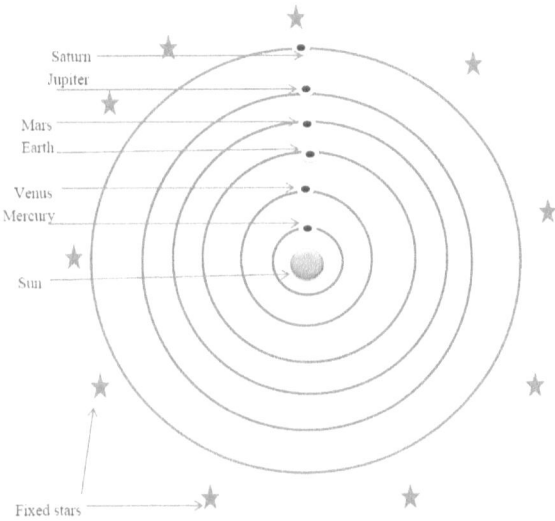

Figure: 1.3. Le model de Copernic

Kepler's three laws can be stated as follows:

First law

The path or orbit of each planet in it motion around the sun is elliptic with the sun at one focus of the ellipse.

Second law

As the planet moves in its orbit, a line drawn from the sun sweep out equal area in equal intervals of time

Third law

The square of the periods of the planets are proportional to the cubes of their mean distance from the sun.

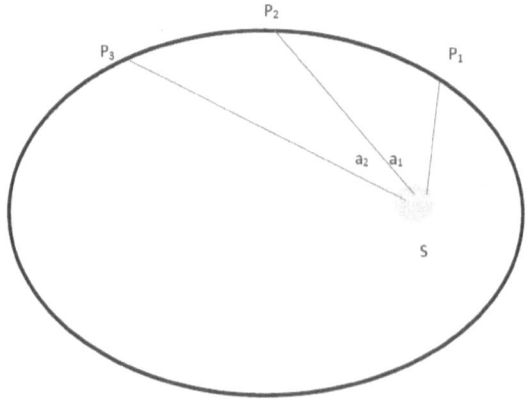

Figure: 1.4. Description of the path of the planet. P_1 P_2, P_3 are the position of the planet. According the Kepler's law, the two area a_1 and a_2 are equal. Area a_1 represents the area SP_1P_2 and area a_2 represents SP_2P_3.

We can obtain the mathematical expression of the third law by using the combine Newton' second law of motion and Newton law of universal gravitation. Thus considering the mass of the sun as M and the mass of the planet as m moving with a velocity $\vec{v}$ and a period of revolution T, the gravitational for applied to the planet will be:

$$F = G\frac{mM}{r^2} \tag{1.12}$$

And the centripetal acceleration of the planet in motion is

$$a = \frac{v^2}{r} = r\omega^2 = \frac{4\pi^2 r}{T^2} \tag{1.13}$$

$$\frac{F}{a} = G\frac{mM}{r^2} \times \frac{T^2}{4\pi^2 r} = \frac{GmM}{4\pi^2}\frac{T^2}{r^3} = m \tag{1.14}$$

So,

$$\frac{T^2}{r^3} = \frac{4\pi^2}{GM} = \text{constant} \qquad (1.15)$$

We can see that T^2 is proportional to r^3

Note that r is the average distance is the average distance between the planet and the sun.

Problem

1. State all Newton three laws of motion and Newton's law of Universal gravitation.

1. Calculate the altitude to which a rocket must be fired in order that the acceleration due to gravity is half of that of the surface of the planet Earth.

2. An airplane traveling at 400mi/hr makes a complete circle in 10min. What is the centripetal acceleration of the airplane? Calculate the centripetal force on the airplane.

3. Calculate the gravitational force between the moon and the planet earth. The mass of the moon is $7.34767309 \times 10^{22} kg$ and the mass of the planet earth is $5.972 \times 10^{24} kg$.

4. Considering the motion of the moon around the planet earth, calculate the velocity of the moon, the centripetal acceleration of the moon toward the planet earth and the centripetal force from the earth on the moon. The distance earth – moon is about 238,900 miles (384,400 km).

Chapter 2: The Planet earth's atmosphere

2.1. Description of the solar system

2.2. Planet Earth atmosphere composition.

2.3. Particles in the troposphere

2.3.1. Sea Salt particles

2.3.2. Particles from fires, rocks, soils and volcanoes

2.1. Description of the solar system.
The planet Earth is in the solar system, a set of eight planet revolving about the sun. Four of them are rocky and terrestrial worlds such as Mercury, Venus, Earth and Mars. The four others are giant gases such as Jupiter, Saturn, Uranus and Neptune. Between the orbits of Mars and Jupiter lies the asteroid belt, which includes the dwarf planet Ceres. Beyond the orbit of Neptune one can observe the disk-shaped Kuiper belt, in which dwarf planet Pluto resides.

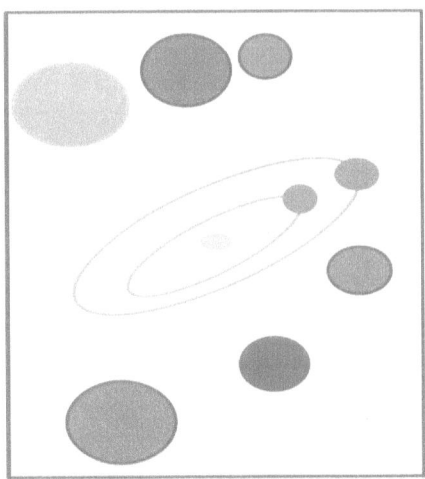

Figure 2.1: Schematic of the Solar system with the eight planets

The sun itself is a huge mass of fire of light of mean radius about 6.96×10^8 m and mass 1.991×10^{30} kg. The planet Earth has a mean radius of 6.38×10^6 m and a mass of 5.98×10^{24} kg. The environment surrounding the planet earth is the atmosphere of the planet earth. It is composed of several gases. Planet earth seen from the above the northern hemisphere rotates counterclockwise about it axis of rotation once in 24hours. It also revolves around the sun in 365 1/4 days.

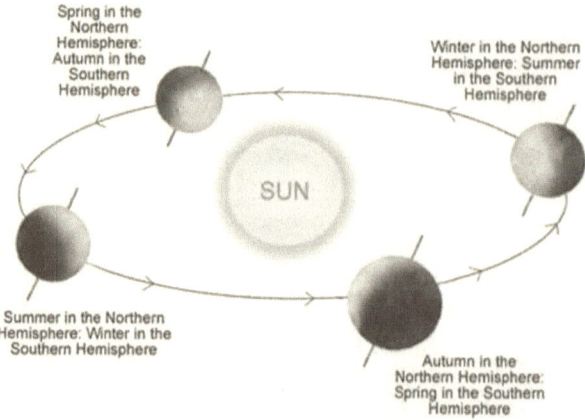

Figure 2.2. Motion of revolution of the planet Earth around the sun

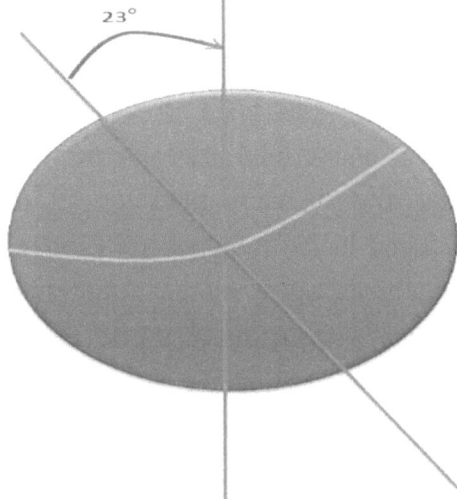

Figure 2.3. Planet rotational motion about its axis

2.2. Planet earth atmosphere composition.

Planet's earth's atmosphere is composed of clouds unequally distributed in the lower 6 to 10 km. 99% of the atmosphere is confined in the lowest 30km while half of the atmosphere is confined in the 6-km layer and that's where most of the clouds are found. Temperature distribution is shown in figure 2.4. The lowest part of atmosphere that is from the surface of the planet Earth up to about 10km is called troposphere. In this region, the temperature decreases with altitude. The word troposphere comes from two words: tropo which stands for turning, and sphere. In other world it stands for changing or turning sphere. The troposphere contains about 80% of the total atmospheric mass. This means that the heaviest particles in the atmosphere are found in the troposphere. Complex atmospheric phenomena occur in the troposphere. Evaporation and heat conduction at the surface of the planet Earth are responsible of the horizontal and the vertical temperature gradient. In the troposphere, it is known that the rising of air cools adiabatically or pseudo-adiabatically resulting in temperature decrease. The size of the troposphere depends on the season and the altitude and the location in the space surrounding the atmosphere. In the tropical regions, it is between 16km and 18km; over the poles, around 8 to 10km in the summer and almost absent in the winter. The troposphere ends with a region called tropopause which is characterized by an increase of static stability and above which resides a statistically stable region called stratosphere which ranges from 20km to 50km. In the stratosphere, the temperature first increases very slowly up to about 20 and 30km, and then increases more rapidly up to 50km where it becomes close the earth's temperature.

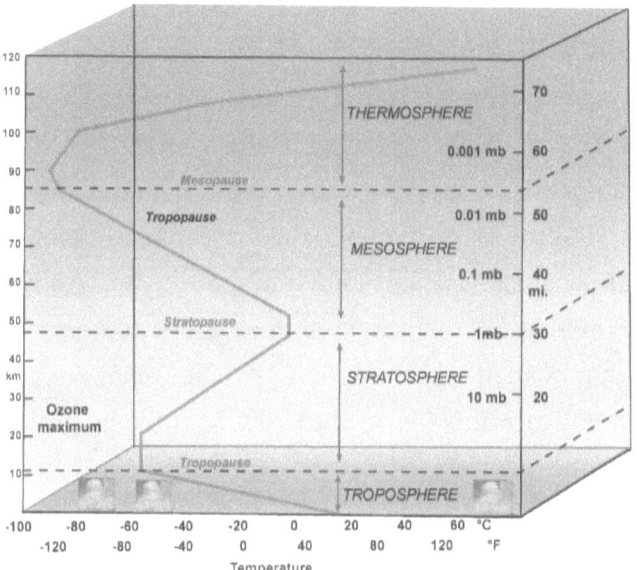

Figure 2.4. Temperature distribution in the earth's atmosphere

The region right after the stratosphere where the temperature is almost constant is called stratopause. Lots of phenomena here are still poorly understood. Above the stratopause, the atmosphere layer is called mesosphere which stands for the middle sphere. In this region, temperature decreases with the height up to around 180K at the height of 85km. Wind speed can also be very height and reach about 150m/s. Temperature distribution in the region is similar to what it is in the troposphere. This can imply that similar atmospheric processes might be happening. Energy source in this region may come from the solar radiation absorption in the stratopause. Above the mesosphere between 80km and 90km, is a region called mesopause where notilucent clouds are sometimes observed. Above the mesopause, which is above 90km, we have a region called thermosphere where temperature increases from 600K to abut 2000K at the height of 500Km. Thermal processes occur in this region because of the sun's activities. Several other phenomena occur in this region such as the ionization, the dissociation of molecules and the diffusion which is more important than mixing; heaviest gases become less concentrated; molecules are dissociated in atoms. Around 100km, atomic oxygen [O] increases with the height while molecules such as nitrogen [N_2] decrease with the height. However, at 100km, atomic oxygen dominates. Later around 500Km to 1000km Helium becomes the more dominant molecule species. Above 1500km, the dominant agent will be atomic hydrogen [H]. The region between 110km and 200km is called thermopause. The region above the thermopause is the exosphere which extends from about 300km to about 500km. In this region, some particles can escape the gravitational field

2.3. Particles in the troposphere

Particles in the atmosphere are also called aerosols. There are four majors groups in the mechanism of formation of aerosols. The first group is the condensation and sublimation of vapor and the formation of anthropogenic source smoke. The second other group concern chemical reaction in the atmosphere. The third is the mechanical disruption and dispersal of matter at the earth's surface such as ocean or mineral. The fourth group is the coagulation of particles which gives larger articles. There's a fifth group which concerns the influx of extraterrestrial particle falling into the earth's atmosphere.

The particles found in the troposphere are both natural and man-made. They are sometimes very difficult to distinguish.

2.3.1. Sea Salt particles

Since ocean makes up about the ¾ of the total surface of the planet earth, we have a huge number of sea particles usually called salt. It is made of sodium chloride (Na^+, Cl^-). There are about 100 sea salt particles per cm^3 in the atmosphere. These particles are widely spread over the atmosphere and participate in some chemical reactions with other particles and trace gases in the atmosphere and extra-terrestrial particles. Pure sodium chloride crystals are hygroscopic and form droplet when the relative humidity is more than 75%. However when the related humidity drops below 75% water vapor in which sea salt is dissolved will evaporate from the sodium chloride until the vapor pressure of the droplet becomes equal to the partial pressure of water vapor in the air. Droplets are mainly form by breaking myriads of air bubbles at the sea surface. Bubbles are produced by breaking small waves, rain or snow falling in water. Near the sea salt, the concentration in salt is mostly high and decrease with distance from the coast and with altitude. But the particles size distribution remains the same. That's why the removal process does not depend on the particles size. The removal process is however less better by sedimentation than by precipitation. The chloride ions in the troposphere are associated with other positive ions to form some giant aerosols in the atmosphere at the vicinity of the oceans.

2.3.2. Particles from fires, rocks, soils and volcanoes

Part of the particles in the troposphere comes from fire, grass. Grass fire can travel very fast and produce clouds. These particles size ranges from 0.1μ in radius for very small particles (Aitkens) to larges particles with 0.1μ in radius, and the giant ones that are more than 0.1μ. There are other particles in the troposphere such as dust from sands or rocks, from dust or sand storms.

Volcanoes also send huge number of particles in the atmosphere. For example the eruption Krakatoa in 1883 in the East Indies produced an explosion of clouds 18 miles high and turn day into night in Batavia, 100 miles away. Another example is the volcano Gunung Agung in Bali in 1963 whose eruption projected particles in the troposphere and in the stratosphere.

Chapter 3
Remote sensing

3.1. Introduction

3.2. History of remote sensing

3.3. Passive and active systems

3.4. Passive sensing

3.4.1. Beer's law and remote sensing of aerosol

3.4.1.1. Lambert's law

3.1. Introduction

Remote sensing is the technique used to collect information on an object without coming into physical contact with that object. For example, how do we get information on the moon without being on the surface of the moon? Even on the surface of the planet earth, remote sensing can be used to obtain information of another object. The technique of remote sensing requires a sensor. Remote sensing is used in many domains such as in meteorology where it is used to profile the temperature of the atmosphere and other particles in the atmosphere which could be oxygen, nitrogen, water vapor. An example is the use of lidar to sense water vapor developed later in this book. Remote sensing can also be used in medical field for detection and therapeutic purposes in the case of tumor for example. It can be used in geology to determine rock, to observe geological faults and tectonic motion. Remote sensing is also used in topography, in agriculture, in oceanography to measure ocean temperature profile and to map ocean currents. The use of remote sensing is becoming very important in the military for explosive detection, combat enemy detection. In this book we are focusing more in the remote sensing in the atmosphere.

3.2. History of remote sensing

Remote sensing can be traced back to the 4th century BC where Aristotle, in the quest of explaining the motion of the planets, and to find a model to that motions used a camera obscura. Camera is the Latin word for room where you house something and obscura is the Latin word for dark, obscure. The instrument was described by Aristotle in his research. However it is possible that this instrument had been used before him. For memory, Aristotle was a Greek philosopher who understood the laws of optics as known in ancient times. The earliest mention of the device of camera obscura was by the Chinese philosopher Mo-Ti on the 5th century BC. He had recorded an inverted image formed by light rays passing through a pinhole in a darkened room. He called this darkened room a "collecting place" or the "locked treasure room." In 1490 Leonardo Da Vinci gave two clear descriptions of the camera obscura in his notebooks. Many of the first camera obscuras were large rooms like the ones illustrated by the Dutch scientist

Reinerus Gemma-Frisius in 1544 for observing solar eclipse. Later the quality of the image was improved by the use of convex lens into the aperture and a mirror to reflect the image down on a surface. The camera obscura" was used by the Johannes Kepler in the early 17th century for astronomical applications. The development of optical theory was significant later during the seventeen century but glass lenses were known much earlier. Major work on remote sensing will really start with the development of photography by Fox Tlbot and Daguerre and Wedgwood, in the earlier nineteenth century. This will be enhanced by the discovery of electromagnetic radiation beyond the visible spectrum by the work of Herschel for the discovery of the infrared radiation, Ritter for the discovery of the ultraviolet radiation and Hertz for the discovery of radio waves. In 1858, the French photographer, Gaspard Felix Tournachon, known also as Nadar, made the first aerial photograph and was a pioneer in artificial lightning in photography. In 1903 air plane was invented and arial photograph from airplanes were soon recorded by 1909. The two World War witness the development and used of aerial photograph to defeat the enemy. So, there no doubt that remote sensing experiences a huge lot of development during World War II and later. Its applications in the medicine as well as in the military and in the atmosphere were beyond measure. Radar was developed in the 1950s and sensors begin to be placed in space. Laser was developed and the moon was interesting human beings. Computer better resolution begins to make possible digital image processing.

3.3. Passive and active systems

In remote sensing, we can distinguish passive systems and active systems. Whether a system is passive or active, it can be imaging or non-imaging. Systems which illuminate the object of study with their own radiation are called active. Systems that sense naturally occurring light such as from solar or any thermal emitted light are called passive. They do not emit the radiation. In addition to these two passive and active systems we can always distinguish system based on the wavelength of radiation they respond to. An imaging system measures the intensity of light reaching it as a function of position on the surface of the planet earth. A two dimensional representation of the intensity can therefore be constructed. A non-imaging system is a system which either does not measure the radiation intensity or does not do so as a function of the position on the earth surface

3.4. Passive sensing
3.4.1. Beer's law and remote sensing of aerosol
3.4.1.1. Lambert's law

Consider a beam of light of frequency γ strike a column of material say a gas. Consider a small part of the material of length, ds. Light intensity is given by Lambert's law.

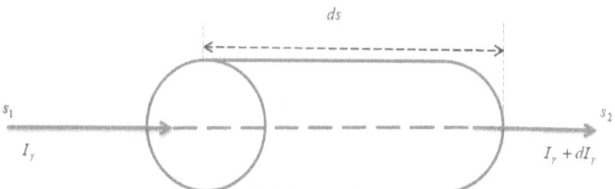

Figure 3.1. Schematic of Lambert's law

Statement of Lambert's law

The change in the intensity along a path ds of a matter is proportional to the amount of matter along the path according to the equation:

$$dI_\gamma = -k_{\gamma,v} I_\gamma ds \qquad (3.1)$$

where $k_{\gamma,v}$ is the volume absorption coefficient.

Integration of equation (3.1) between s_1 and s_2 gives a general solution to Lambert's law:

$$I_\gamma(s_2) = I_\gamma(s_1).T_\gamma(s_1,s_2) \qquad (3.2)$$

This law is used in the determination of ozone part-integrated and water vapor concentrations.

Most often one can idealize the atmosphere as a horizontal stratified medium and define the path relative to the vertical.

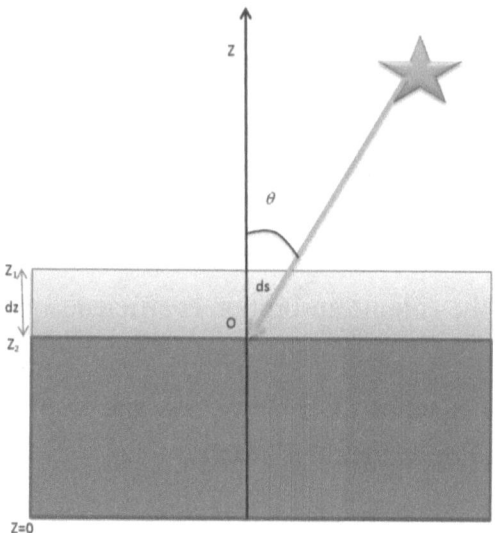

Figure 3.2. A sun ray travelling through the atmosphere under the zenith angle θ.

The sun ray arrives and strikes a portion of the atmosphere under the zenith angle θ

The transmission along the path from the vertical by the angle θ is related to the transmission along the vertical path according to:

$$T_\gamma(s_1, s_2) = \exp\left(-\int_{s_1}^{s_2} k_{\gamma,v}\, ds\right) \quad (3.3)$$

Here $\cos(\theta) = \dfrac{dz}{ds} = \mu$

Then $T_\gamma(s_1, s_2) = \exp\left(-\dfrac{1}{\mu}\int_{s_1}^{s_2} k_{\gamma,v}\, dz\right)$

Or

$$T_\gamma(s_1, s_2) = \exp\left(-\dfrac{\tau_\gamma(z_1, z_2)}{\mu}\right) \quad (3.4)$$

The optical path $\tau_\gamma(z_1, z_2)$ is now measured along the vertical and it is referred to as the optical depth. The optical depth can be written in terms of the mass absorption coefficient, $k_{\gamma,m}$ as follows:

$$\tau(s_1, s_2) = \int_{s_1}^{s_2} \rho_a k_{\gamma,m}\, dz \quad (3.5)$$

So

$$T_\gamma(s_1, s_2) = \exp\left(-\mu^{-1}\int_{z_1}^{z_2} \rho_a k_{\gamma,m}\, dz\right) \quad (3.6)$$

We can define the optical mass as:

$$\tau(z_1, z_2) = \int_{z_1}^{z_2} \rho_a \, dz \qquad (3.7)$$

The SI unit of the optical mass will be kg/m^2

Chapter 4: Background on Maxwell's laws in electromagnetism

4.1. Maxwell equations

4.2. Equation of conservation of charge

4.3. Electromagnetic energy

4.3.1. Electrostatics energy

4.3.2. Magnetic energy

4.3.3. The Poynting vector

4.3.4. The total energy in a linear medium

4.4. Energy conservation

4.5. The Electromagnetic wave equations

4.6. Electromagnetism wave equations in an inhomogeneous medium

4.7. Poisson and Laplace equations

4.8. Electromagnetic radiation

4.1. Maxwell equations

Maxwell's equations are a series of four very important equations of electromagnetic field wave properties that can be found in various places in a Maxwell 1961 paper on physical lines of force. They express how electric charges produce electric fields (Gauss' law for electricity), the experimental absence of magnetic monopoles (Gauss' law for magnetism), how electric current and change in electric fields can produce magnetic fields (Ampere's circuital law), and how magnetic field and change in magnetic fields can produce electric fields (Faraday's law of induction).

For convenience, we recapitulate them here in MKSA unit system (Meter for length, Kilogram for mass, Second for time, and Ampere for current intensity), which is adopted in SI (Système International).

Maxwell equations in vacuum and in a homogeneous medium: differential form

Gauss' law for electricity

$$div\,\vec{E} = 0 \text{ or } \vec{\nabla}\cdot\vec{E} = 0 \tag{4.1}$$

Gauss' law for magnetism

$$\text{div}\,\vec{B}=0 \quad \text{or} \quad \vec{\nabla}\cdot\vec{B}=0 \qquad (4.2)$$

Ampere's circuital law

$$\text{curl}\,\vec{B}=\varepsilon_0\mu_0\frac{\partial \vec{E}}{\partial t} \quad \text{or} \quad \vec{\nabla}\times\vec{B}=\varepsilon_0\mu_0\frac{\partial \vec{E}}{\partial t} \qquad (4.3)$$

Faraday's law for induction

$$\text{curl}\,\vec{E}=-\frac{\partial \vec{B}}{\partial t} \quad \text{or} \quad \vec{\nabla}\times\vec{E}=-\frac{\partial \vec{B}}{\partial t} \qquad (4.4)$$

Maxwell equations in inhomogeneous medium: differential form

1. Gauss' law for electricity

$$\text{div}\,\vec{D}=\rho \quad \text{or} \quad \vec{\nabla}\cdot\vec{D}=\rho \qquad (4.5)$$

2. Gauss' law for magnetism (absence of monopole)

$$\text{div}\,\vec{B}=0 \text{ or } \vec{\nabla}\cdot\vec{B}=0 \qquad (4.6)$$

3. Ampere's circuital law

$$\text{curl}\,\vec{H}=\vec{J}+\frac{\partial \vec{D}}{\partial t} \text{ or } \vec{\nabla}\times\vec{H}=\vec{J}+\frac{\partial \vec{D}}{\partial t} \qquad (4.7)$$

Faraday's law of induction

$$\text{curl}\,\vec{E}=-\frac{\partial \vec{B}}{\partial t} \quad \text{or} \quad \vec{\nabla}\times\vec{E}=-\frac{\partial \vec{B}}{\partial t} \qquad (4.8)$$

Maxwell equations in inhomogeneous medium: integral form

1. Gauss' law for electricity

$$\oint_S \vec{D}\cdot d\vec{A}=q \qquad (4.9)$$

2. Gauss' law for magnetism (absence of monopole)

$$\oint_S \vec{B} \cdot \vec{dA} = 0 \qquad (4.10)$$

3. Faraday's law of induction

$$\oint_C \vec{E} \cdot \vec{dl} = -\int_S \frac{\partial \vec{B}}{\partial t} \cdot \vec{dA} \qquad (4.11)$$

4. Ampere's circuital law

$$\oint_C \vec{H} \cdot \vec{dl} = \int_S \vec{J} \cdot \vec{dA} + \int_S \frac{\partial \vec{D}}{\partial t} \cdot \vec{dA} \qquad (4.12)$$

$k = \dfrac{1}{4\pi\varepsilon_0}$ is coulomb's constant and

$$c^2 = \frac{1}{\mu_0 \varepsilon_0} \qquad (4.13)$$

$\vec{H}$ is the magnetic field strength vector of magnetic field in A/m

$\vec{E}$ is the electric field in V/m or N/C

$\vec{D}$ is the electric displacement field in C/m^2

$\vec{J}$ is the free current density in A/m²

$d\vec{A}$ is the differential vector element of surface area $\vec{A}$ in m²

$\vec{P}$ is the polarization density vector in C/m²

$\vec{M}$ is the magnetization density vector in A/m

μ is the magnetic permeability in Henry/m or N/A²

In a linear material

$$\vec{P} = \chi_e \varepsilon_0 \vec{E} \tag{4.14}$$

$$\vec{M} = \chi_m \vec{H} \tag{4.15}$$

$$\vec{D} = \varepsilon_0 \vec{E} + \vec{P} = (1 + \chi_e) \varepsilon_0 \vec{E} = \varepsilon \vec{E} \tag{4.16}$$

$$\vec{B} = \mu_0 (\vec{H} + \vec{M}) = (1 + \chi_m) \mu_0 \vec{H} = \mu \vec{H} \tag{4.17}$$

with the following definitions:

χ_e is the electric susceptibility of the material

χ_m is the magnetic susceptibility of the material
ε is the electric permittivity of the material
μ is the magnetic permeability of the material
In the vacuum $\varepsilon = \varepsilon_0$ and $\mu = \mu_0$
$\mu_0 = 4\pi \times 10^{-7}\ H/m$ (Henry / meter)
$\varepsilon_0 = 8.85419 \times 10^{-12}\ Farads/m$

4.2. Equation of conservation of charge

Let us take the divergence of the curl of $\vec{B}$

$$\vec{\nabla}(curl\ \vec{B}) = 0 = \vec{\nabla}\cdot(\vec{\nabla}\times\vec{B}) = \vec{\nabla}\cdot(\mu\vec{J} + \mu\frac{\partial \vec{D}}{\partial t}) \quad (4.18)$$

Therefore

$$\vec{\nabla}\vec{J} + \frac{\partial(\vec{\nabla}\cdot\vec{D})}{\partial t} = 0 \text{ or } \vec{\nabla}\cdot\vec{J} + \frac{\partial \rho}{\partial t} = 0 \quad (4.19)$$

4.3. Electromagnetic energy

The electromagnetic energy is the contribution of the electrostatics energy and the magnetic energy. We would consider these two energies separately and then add them up have the form of the electromagnetic energy

4.3.1. Electrostatics energy

The electrostatics energy in linear dielectrics of constant of permittivity ε can be written:

$$U_E = \frac{1}{2}\int_{Volume} \vec{D}\cdot\vec{E}\,dv \qquad (4.20)$$

The integration is over the volume of the system external to the conductor. It can be of course extended to all space included inside the conductor where the electric field $\vec{E}$ is zero. To the question of where is the electrostatic energy located, it is easy to imagine that it is stored inside the electric field itself.

The electrostatic energy density or electrostatic energy per unit of volume can then be defined as

$$u_E = \frac{dU_E}{dv} = \frac{1}{2}\vec{D}\cdot\vec{E} = \frac{1}{2}\frac{\vec{D}^2}{\varepsilon} \qquad (4.21)$$

4.3.2. Magnetic energy

The formulation of the magnetic energy in terms of the vectors $\vec{B}$ and $\vec{H}$ is of considerable interest because it provides picture of how the magnetic energy is stored in the magnetic field itself. The magnetic energy can be written as:

$$U_M = \frac{1}{2}\int_{Volume} \vec{H}\cdot\vec{B}\, dv \qquad (4.22)$$

The previous expression is analogous to the one obtained for the electrostatic energy. It is important to note that it is restricted to systems containing linear magnetic media.

Similarly, the magnetic energy density or magnetic energy per unit of volume in isotropic, linear and magnetic materials can be written as follows:

$$u_M = \frac{dU_M}{dv} = \frac{1}{2}\vec{H}\cdot\vec{B} = \frac{1}{2}\frac{\vec{B}^2}{\mu} \qquad (4.23)$$

4.3.3. The Poynting vector

The Poynting vector $\vec{S}$ can be defined as the local energy flow per unit time per unit area. It is named after John Henry Poynting (1852-1914) who discovered it, and represents the energy flux per unit of area in (W/m^2) of an electromagnetic field.

$$\vec{S} = \vec{E} \times \vec{H} \qquad (4.24)$$

4.3.4. The total energy in a linear medium

The total energy in a linear medium is the sum of the electrostatic and magnetic energies:

$$U = U_E + U_M \qquad (4.25)$$

The total electromagnetic energy density can be written as follows:

$$u = u_E + u_M = \frac{1}{2}(\vec{E} \cdot \vec{D} + \vec{H} \cdot \vec{B}) \qquad (4.26)$$

4.4. Energy conservation

The expression that shows the electromagnetic energy conservation law is given by:

$$\vec{\nabla} \cdot \vec{S} + \frac{\partial u}{\partial t} = -\vec{J} \cdot \vec{E}. \quad (4.27)$$

In this expression the term $\vec{J} \cdot \vec{E}$ represents the work done by the local field on charged particles per unit volume. The energy conservation relation can be proved from Maxwell's equations.

4.5. The Electromagnetic wave equations

For the magnetic field

From Maxwell equation in vacuum we know

$$\vec{\nabla} \times \vec{B} = \varepsilon_0 \mu_0 \frac{\partial \vec{E}}{\partial t} \quad (4.28)$$

By taking the rotational or the curl of both sides, we have

$$\vec{\nabla} \times (\vec{\nabla} \times \vec{B}) = \varepsilon_0 \mu_0 \frac{\partial(\vec{\nabla} \times \vec{E})}{\partial t} \quad \text{or}$$

$$\vec{\nabla}\, \vec{\nabla} \cdot \vec{B} - \vec{\nabla} \cdot \vec{\nabla}\, \vec{B} = \varepsilon_0 \mu_0 \frac{\partial(\vec{\nabla} \times \vec{E})}{\partial t}. \qquad (4.29)$$

But, one of the Maxwell equations' properties is that $\vec{\nabla} \cdot \vec{B} = 0$ and

$\vec{\nabla} \times \vec{E} = -\frac{\partial \vec{B}}{\partial t}$. Then equation (33) becomes:

$$\vec{\nabla} \cdot \vec{\nabla}\, \vec{B} = \varepsilon_0 \mu_0 \frac{\partial^2 \vec{B}}{\partial t^2} \quad \text{or} \quad \vec{\nabla} \cdot \vec{\nabla}\, \vec{B} = \frac{1}{c^2} \frac{\partial^2 \vec{B}}{\partial t^2}. \qquad (4.30)$$

The previous equation is the wave equation for magnetic field $\vec{B}$ in vacuum. It involves the scalar Laplacian $\Delta = \vec{\nabla} \cdot \vec{\nabla}$ of the magnetic field $\vec{B}$.

For the electric field

From Maxwell's equation in vacuum we know

$$\vec{\nabla} \times \vec{B} = \varepsilon_0 \mu_0 \frac{\partial \vec{E}}{\partial t}$$

Let's take now time differentiation of both side of the previous equality. We then have

$$\vec{\nabla} \times \frac{\partial \vec{B}}{\partial t} = \varepsilon_0 \mu_0 \frac{\partial^2 \vec{E}}{\partial t^2}$$

and using $\vec{\nabla} \times \vec{E} = -\frac{\partial \vec{B}}{\partial t}$, we

obtain $\vec{\nabla} \times (\vec{\nabla} \times \vec{E}) = -\varepsilon_0 \mu_0 \frac{\partial^2 \vec{E}}{\partial t^2}$ or

$$\vec{\nabla}\vec{\nabla} \cdot \vec{E} - \vec{\nabla} \cdot \vec{\nabla} \vec{E} = -\varepsilon_0 \mu_0 \frac{\partial^2 \vec{E}}{\partial t^2} \qquad (4.31)$$

But one of the Maxwell's equations properties in vacuum is that

$\vec{\nabla} \cdot \vec{E} = 0$; therefore the wave equation for the Electric field $\vec{E}$ in vacuum becomes:

$$\vec{\nabla}\cdot\vec{\nabla}\vec{E}=\varepsilon_0\mu_0\frac{\partial^2\vec{E}}{\partial t^2} \text{ or } \vec{\nabla}\cdot\vec{\nabla}\vec{E}=\frac{1}{c^2}\frac{\partial^2\vec{E}}{\partial t^2} \quad (4.32)$$

This Equation involves the scalar Laplacian $\Delta=\vec{\nabla}\cdot\vec{\nabla}$ of the electric field $\vec{E}$.

4.6. Electromagnetism wave equations in an inhomogeneous medium

In the case that we consider both the scalar potential φ and the vector potential $\vec{A}$ so that the vector electric field can be written

$$\vec{E}=-\vec{\nabla}\varphi-\frac{\partial\vec{A}}{\partial t} \quad (4.33)$$

with $\vec{B}=\vec{\nabla}\times\vec{A}$

and the inhomogeneous Maxwell's equations:
Gauss' law $\vec{\nabla}\cdot\vec{E}=\frac{\rho}{\varepsilon_0}$ and Oersted's law

$\vec{\nabla}\times\vec{B}-\frac{1}{c^2}\frac{\partial\vec{E}}{\partial t}=\mu_0\vec{J}$, the previous equation leads to the wave equation for the potentials φ and $\vec{A}$ provided that we have the constraint:

$$\frac{1}{c^2}\frac{\partial \varphi}{\partial t}+\vec{\nabla}\cdot\vec{A}=0 \qquad (4.34)$$

The above constraint which consists of fixing the divergence of the vector potential is called Lorentz gauge.

Substituting our electric field into Gauss' law yields

$$\frac{\rho}{\varepsilon_0}=\vec{\nabla}\cdot\vec{E}=-\vec{\nabla}^2\varphi-\frac{\partial}{\partial t}(\vec{\nabla}\cdot\vec{A})=-\vec{\nabla}^2\varphi+\frac{1}{c^2}\frac{\partial^2 \varphi}{\partial t^2} \qquad (4.35)$$

which is the wave equation for electric potential. Also, substituting $\vec{B}=\vec{\nabla}\times\vec{A}$ into Oersted's law and using Lorentz gauge condition we can write the wave equation for the vector potential:

$$-\vec{\nabla}^2\vec{A}+\frac{1}{c^2}\frac{\partial^2 \varphi}{\partial t^2}=\mu_0\vec{J} \qquad (4.36)$$

4.7. Poisson and Laplace equations

Let consider the differentiation form of Gauss law for electricity:

$\vec{\nabla} \cdot \vec{D} = \rho$ with the electric displacement field $\vec{D} = \varepsilon_0 \vec{E}$

Therefore we have $\vec{\nabla} \cdot \vec{E} = \dfrac{\rho}{\varepsilon_0}$. Also the electrostatic field derived from a potential φ that is

$\vec{E} = -\vec{\nabla}\varphi$ (Electric field is the opposite of the gradient of a potential). Combining the two previous expressions we have:

$$\vec{\nabla}^2 \varphi = -\dfrac{\rho}{\varepsilon_0} \qquad (4.37)$$

The scalar differential operator $\vec{\nabla}^2$ can also be written Δ and is called the Laplacian.

Equation (4.37) is called Poisson's Equation.

4.8. Electromagnetic radiation

The electromagnetic spectrum is the distribution of electromagnetic radiation according to energy, frequency or wavelength.

The infrared radiation is usually subdivided into three basic segments: the Near infrared radiation (0.076–1.5 micrometers), the middle infrared radiation (1.5–5.6 micrometers and the far infrared radiation (5.6 micrometers–1 mm).

The electromagnetic radiation as seen in figure 4.1 is wide, ranging from the Gamma rays to shortwave and the radio waves.

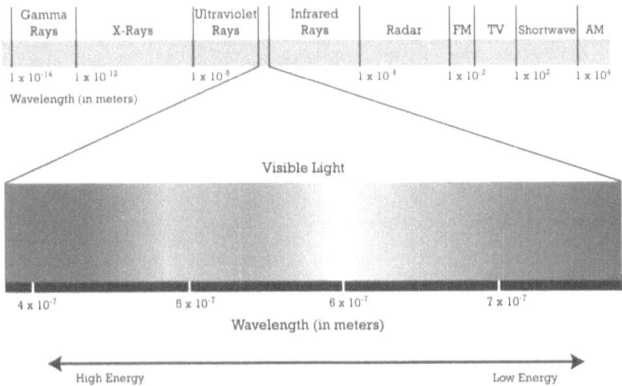

Figure 4.1. Electromagnetic spectrum

Chapter 5

Light polarization in a given medium

5.1. State of polarization

5.2. The linear polarization

5.3. Circular polarization

5.4. Elliptical polarization

5.5. Natural and partially polarized light

5.6. Matrix representation of polarization

5.7. Jones vector for linearly polarized light

5.8. Jones vector for circular polarization

5.9. Jones matrix for a rotator

5.1. State of polarization

Consider a plane monochromatic electromagnetic wave defined by the following electric and magnetic fields:

$$\vec{E}(\vec{r},t) = \vec{E}_0 \, e^{i(kz-\omega t)} \; ; \; \vec{B}(\vec{r},t) = \vec{B}_0 \, e^{i(kz-\omega t)} \quad (5.1)$$

We assume here that E and B are propagating in an isotropic medium and the positive z-direction. The amplitudes $\vec{E}_0$ and $\vec{B}_0$ of the electric and magnetic fields $\vec{E}$ and $\vec{B}$ are complex vectors and determine the state of polarization of the electromagnetic wave. The electric field is in the x-y plane and its plane of vibration is the x-z plane. If we consider $(\vec{i}, \vec{j})$ as the orthonormal base in the x-y plane, then we can write:

$$\vec{E}(\vec{r},t) = \vec{E}(x,y,z,t) = E_x(z,t)\vec{i} + E_y(z,t)\vec{j}. \quad (5.2)$$

$\vec{E}(\vec{r},t)$ can also be written:

$$\begin{aligned}\vec{E}(\vec{r},t) &= \vec{E}_0(z,t)e^{i(kz-\omega t)} \\ &= (E_{0x}\vec{i} + E_{0y}(z,t)\vec{j})e^{i(kz-\omega t)}\end{aligned} \quad (5.3)$$

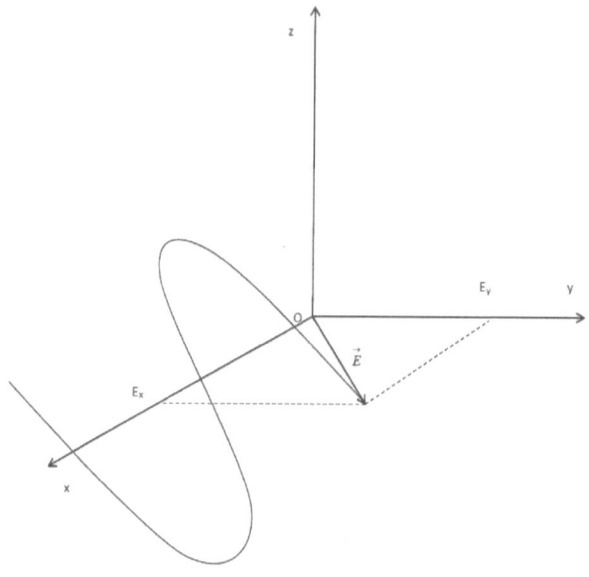

5.2. The linear polarization

Light is linearly polarized if the both components E_x and E_y of the electric field oscillate in phase (0°) or out of phase (180°). A linearly polarized wave is also called plane-polarized or p-state light because the electric field vectors are distributed sinusoidally along the direction of propagation.

Example 1

Describe the wave given by the electric field:

$$\vec{E}(\vec{r},t) = E_o \sin\left(\frac{2\pi}{\lambda}x - \omega t\right)\vec{j} - E_o \sin\left(\frac{2\pi}{\lambda}x - \omega t\right)\vec{k} \quad (5.4)$$

where $\vec{j}$ and $\vec{k}$ are unit vectors in the Cartesian coordinates system. What is the scalar amplitude of $\vec{E}$?

Solution

We can rewrite the wave as follows:

$$\vec{E}(\vec{r},t) = (E_o \vec{j} - E_o \vec{k})\sin\left(\frac{2\pi}{\lambda}x - \omega t\right) \quad (5.5)$$

This wave travels in the x-positive direction with a constant amplitude of $\vec{E}_o = E_o \vec{j} - E_o \vec{k}$. Also, we can write the two components of the wave:

$$E_y = E_o \sin\left(\frac{2\pi}{\lambda}x - \omega t\right) \quad (5.6)$$

$$E_z = -E_o \sin\left(\frac{2\pi}{\lambda}x - \omega t\right) \quad (5.7)$$

We notice that E_y and E_z are related by a linear relation: $E_y = -E_z$. The wave is then linearly polarized in the x-direction. The plane of polarization is tilted at $90°+45° = 135°$ to the x-y plane.

Amplitude of the wave

The amplitude of the wave is
$$E = \sqrt{(E_0)^2 + (-E_0)^2} = E_0\sqrt{2}$$

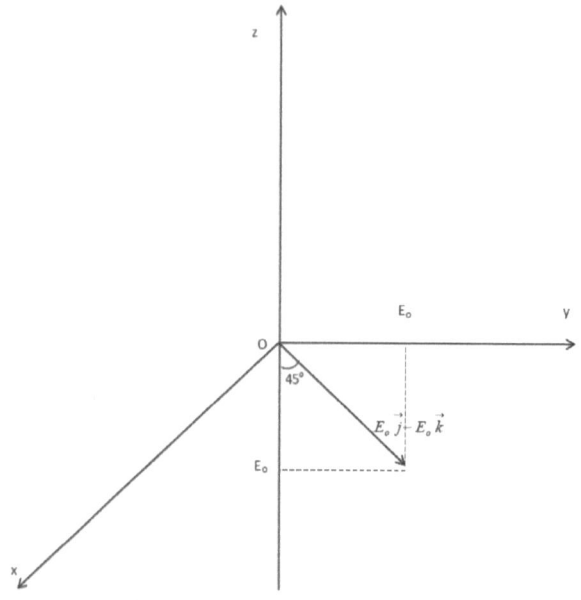

Example 2

a. Describe the wave given by the superposition of the disturbances:

$$\vec{E}_x(y,t) = \vec{i}\, E_0 \cos k(y-vt)$$

$$\vec{E}_z(y,t) = -\vec{k} E_0 \cos k(y-vt)$$

b. Calculate $\vec{E}(0,t)$ and $\vec{E}(0,0)$

Solution

a. The superposition of the two waves gives:

$$\vec{E}(y,t) = (\vec{i} E_0 - \vec{k} E_0) \cos k(y-vt)$$

The amplitude of the superposition is constant. Also, $E_z = -E_x$. Therefore the wave is planar and linearly polarized.

b. $\vec{E}(0,t) = (\vec{i} E_0 - \vec{k} E_0) \cos(kvt)$ and

$$\vec{E}(0,0) = \vec{i} E_0 - \vec{k} E_0$$

5.3. Circular polarization

Suppose the two perpendicular electric fields are written

$$\vec{E}(z,t) = \vec{i} E_0 \cos(kz - \omega t) + \vec{j} E_0 \sin(kz - \omega t) \quad (5.8)$$

Therefore

$$\begin{cases} E_x = E_0 \cos(kz - \omega t) \\ E_y = E_0 \sin(kz - \omega t) \end{cases} \quad (5.9)$$

Using a trigonometric rule, we can find that:

$$\left(\frac{E_x}{E_o}\right)^2 + \left(\frac{E_y}{E_o}\right)^2 = 1 \quad (5.10)$$

The electric filed $\vec{E}$ rotates clockwise with constant amplitude. The endpoint of the vector field $\vec{E}$ sweeps out a circle. This type of field is said to be right circularly polarized and is referred to as R-state wave. However, if the cosine in $\vec{E}$ is shifted into a negative sine such as $\vec{E}(z,t) = \vec{i} E_0 \cos(kz - \omega t) - \vec{j} E_0 \sin(kz - \omega t)$, the electric field has a constant magnitude but now rotates counterclockwise. The field is said to be left circularly polarized or L-state

Example

What is the difference between the R-state wave

$\vec{E} = E_0[\vec{i}\cos(kx-\omega t) + \vec{j}\sin(kx-\omega t)]$ and the wave

$\vec{E}' = E_0[\vec{i}\sin(kx-\omega t) + \vec{j}\cos(kx-\omega t)]$?

Solution

The two waves have the same amplitude:

$\left\|\vec{E_0}\right\|^2 = E_0^2$ and $\left\|\vec{E_0}'\right\|^2 = E_0^2$

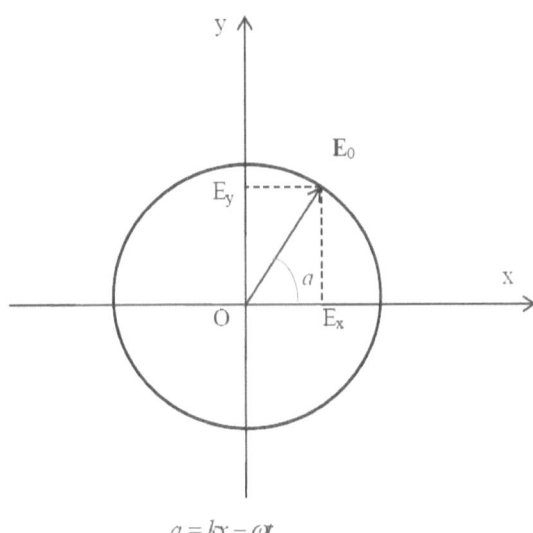

$a = kx - \omega t$

$$\begin{cases} E_x = E_0 \cos(kz - \omega t) \\ E_y = E_0 \sin(kz - \omega t) \end{cases} \quad (5.11)$$

$\left(\dfrac{E_x}{E_o}\right)^2 + \left(\dfrac{E_y}{E_o}\right)^2 = 1$. Therefore $\vec{E}$ is right circularly polarized. The end of vector $\vec{E}$ rotates clockwise when time changes. However, for $\vec{E}'$ we have:

$\begin{cases} E_x' = E_0 \sin(kz - \omega t) \\ E_y' = E_0 \cos(kz - \omega t) \end{cases}$, and $\left(\dfrac{E_x'}{E_o}\right)^2 + \left(\dfrac{E_y'}{E_o}\right)^2 = 1$

which also implies a circular wave. But contrarily to $\vec{E}$, the endpoint of $\vec{E}'$ sweeps the circle counterclockwise. The wave is left circularly polarized (L-state).

5.4. Elliptical polarization

We will use a more general expression of $E_x(z,t)$ and $E_y(z,t)$:

$$\begin{cases} E_x = E_{0x} \cos(kz - \omega t) \\ E_y = E_{0y} \cos(kz - \omega t + \varepsilon) \end{cases} \quad (5.12)$$

where ε is the relative phase

We recall that linear polarization occurs for $\varepsilon = 0$ or is an integral multiple of $\pm 2\pi$. Circular polarization occurs when the relative phase $\varepsilon = -\dfrac{\pi}{2} + 2m\pi$, with $m = 0, \pm 1, \pm 2, \pm 3, \ldots$.

For the elliptical or ε-state light, the end of the electrical field, $\vec{E}$, sweeps out an ellipse or more precisely an elliptical helix. We can try to isolate ε by using trigonometric rules. Also, the expansion of E_y gives:

$$E_y = E_{0y}\cos(kz-\omega t)\cos(\varepsilon) - \sin(kz-\omega t)\sin(\varepsilon) \quad (5.13)$$

$$\dfrac{E_y}{E_{0y}} = E_x \cos(\varepsilon) - \sin(kz-\omega t)\sin(\varepsilon) \quad (5.14)$$

$$\dfrac{E_y}{E_{0y}} = \cos(\varepsilon) - \sin(kz-\omega t)\sin(\varepsilon) \quad (5.15)$$

We finally have:

$$\begin{cases} \cos(kz-\omega t) = \dfrac{E_x}{E_{ox}} \\ \sin(kz-\omega t) = -\dfrac{1}{\sin(\varepsilon)}\left[\dfrac{E_y}{E_{oy}} - \dfrac{E_x}{E_{ox}}\cos(\varepsilon)\right] \end{cases} \quad (5.16)$$

and

$$\left(\dfrac{E_x}{E_{ox}}\right)^2 + \dfrac{1}{\sin^2(\varepsilon)}\left[\dfrac{E_y}{E_{oy}} - \dfrac{E_x}{E_{ox}}\cos(\varepsilon)\right]^2 = 1 \quad (5.17)$$

$$\left(\dfrac{E_x}{E_{ox}}\right)^2 \sin^2(\varepsilon) + \left(\dfrac{E_y}{E_{oy}}\right)^2 - 2\left(\dfrac{E_x}{E_{ox}}\right)\left(\dfrac{E_y}{E_{oy}}\right)\cos(\varepsilon)$$
$$+ \left(\dfrac{E_x}{E_{ox}}\right)^2 \cos^2(\varepsilon) = \sin^2(\varepsilon) \quad (5.18)$$

which gives:

$$\left(\dfrac{E_x}{E_{ox}}\right)^2 + \left(\dfrac{E_y}{E_{oy}}\right)^2 - 2\left(\dfrac{E_x}{E_{ox}}\right)\left(\dfrac{E_y}{E_{oy}}\right)\cos(\varepsilon) = \sin^2(\varepsilon) \quad (5.19)$$

Equation (2.15) is the equation of an ellipse tilted at an angle α to the E_x axis such that:

$$\tan(2\alpha) = \left(\dfrac{2E_{ox}E_{oy}}{E^2_{ox} - E^2_{oy}}\right)\cos(\varepsilon) \quad (5.20)$$

Example

Describe the state of polarization of the wave characterized by the following equation:

$$\begin{cases} E_x = E_0 \cos(kx - \omega t) \\ E_y = E_0 \sin(kx - \omega t + \dfrac{\pi}{4}) \end{cases}$$

(5.21)

Solution

Let us expand E_y:

$$E_y = E_0 \left[\sin(kx - \omega t)\cos(\frac{\pi}{4}) + \cos(kx - \omega t)\sin(\frac{\pi}{4}) \right] \quad (5.22)$$

$$E_y = \frac{\sqrt{2}}{2} E_0 \left[\sin(kx - \omega t) + \cos(kx - \omega t) \right]$$

$$E_y = \frac{\sqrt{2}}{2} E_0 \left[\sin(kx - \omega t) + \frac{E_x}{E_0} \right] \quad (5.23)$$

Then the system becomes:

$$\begin{cases} \cos(kx-\omega t) = \dfrac{E_x}{E_0} \\ \sin(kx-\omega t) = \dfrac{2}{\sqrt{2}E_0}\left(E_y - \dfrac{\sqrt{2}}{2}E_x\right) \end{cases} \quad (5.24)$$

which implies:

$$\left(\dfrac{E_x}{E_0}\right)^2 + \dfrac{2}{E_0^2}\left(E_y - \dfrac{\sqrt{2}}{2}E_x\right)^2 = 1 \quad (5.25)$$

After further expansions, the following equation can be obtained:

$$\left(\dfrac{E_x}{E_0}\right)^2 + \left(\dfrac{E_y}{E_0}\right)^2 - 2\left(\dfrac{E_x}{E_0}\right)\left(\dfrac{E_y}{E_0}\right)\cos(\varepsilon) = \sin^2(\varepsilon) \quad (2.26)$$

with $\cos(\varepsilon) = \cos(\dfrac{\pi}{4}) = \sin(\dfrac{\pi}{4}) = \dfrac{\sqrt{2}}{2}$

Equation (2.26) describes an ellipse at the angle α to the E_x axis such that:

$$\tan(2\alpha) = \dfrac{2E_{0x}E_{0y}}{E_{0x}^2 - E_{0y}^2}\cos(\varepsilon) \quad (5.27)$$

But because $E_{0x} = E_{0y} = E_0$, $\tan(2\alpha)$ tends to ∞ and α tends to $\frac{\pi}{4}$.

5.5. Natural and partially polarized light

A natural light is composed of both partially polarized and unpolarized light. If I_u is the flux density of the polarized component and I_p the flux density of the unpolarized component, the irradiance is defined as $I_u + I_p$ and the fractional polarized component or degree of polarization α is defined as:

$$\alpha = \frac{I_p}{I_u + I_p} \qquad (5.28)$$

5.6. Matrix representation of polarization

It is interesting to represent the state of polarization by matrices. The x-component and the y-component of the electromagnetic field E can be written as complex numbers by Jones column vector:

$$E_x = E_{0x} \exp(i(kz - \omega t + \phi_x));$$
$$E_y = E_{0y} \exp(i(kz - \omega t + \phi_y)). \quad (5.29)$$

For a plane polarized light $\phi_y - \phi_x = 0, \pi$

By combining the components we obtain

$$\vec{E} = \hat{i} E_{ox} e^{i(kz-\omega t+\varphi_x)} + \hat{j} E_{oy} e^{i(kz-\omega t+\varphi_y)} \quad (5.30)$$

which can be written

$$\vec{E} = \left[\hat{i} E_{ox} e^{i\varphi_x} + \hat{j} E_{oy} e^{i\varphi_y}\right] e^{i(kz-\omega t)} \quad (5.31)$$

Or

$$\vec{E} = \tilde{E}_o e^{i(kz-\omega t)} \quad (5.32)$$

The terms in brackets represents the complex amplitude of the plane wave

The state of polarization of light is determined by the relative amplitudes (E_{ox}, E_{oy}) and, the relative phases ($\delta = \varphi_y - \varphi_x$) of these components.

The complex amplitude is written as a two-element matrix, the Jones vector

$$\tilde{E}_o = \begin{bmatrix} \tilde{E}_{ox} \\ \tilde{E}_{oy} \end{bmatrix} = \begin{bmatrix} E_{ox}e^{i\varphi_x} \\ E_{oy}e^{i\varphi_y} \end{bmatrix} = e^{i\varphi_x} \begin{bmatrix} E_{ox} \\ E_{oy}e^{i\delta} \end{bmatrix} \qquad (5.33)$$

Let us represents Jones vector when the horizontally polarized light. In this case the electric field oscillations are only along the x-axis. As such, the Jones vector is written,

$$\tilde{E}_o = \begin{bmatrix} \tilde{E}_{ox} \\ \tilde{E}_{oy} \end{bmatrix} = \begin{bmatrix} E_{ox}e^{i\varphi_x} \\ 0 \end{bmatrix} = \begin{bmatrix} A \\ 0 \end{bmatrix} = A \begin{bmatrix} 1 \\ 0 \end{bmatrix} \qquad (5.34)$$

where we have set the phase, $\varphi_x = 0$, for convenience

with the normalized form being $\begin{bmatrix} 1 \\ 0 \end{bmatrix}$

Let us represents Jones vector when the horizontally polarized light. In this case the electric field oscillations are only along the y-axis. As such, the Jones vector is written

$$\tilde{E}_o = \begin{bmatrix} \tilde{E}_{ox} \\ \tilde{E}_{oy} \end{bmatrix} = \begin{bmatrix} 0 \\ E_{oy}e^{i\varphi_y} \end{bmatrix} = \begin{bmatrix} 0 \\ A \end{bmatrix} = A\begin{bmatrix} 0 \\ 1 \end{bmatrix} \quad (5.35)$$

where we have set the phase $\varphi_y = 0$, for convenience

5.7. Jones vector for linearly polarized light

Consider a linearly polarized light at an arbitrary angle. If the phases are integer multiple of π such that $\delta = m\pi$, with m = 0, ±1, ±2, ±3, ..., then we have,

$$\frac{E_x}{E_y} = (-1)^m \frac{E_{ox}}{E_{oy}} \quad (5.36)$$

and the Jones vector is simply a line inclined at an angle $\alpha = \tan^{-1}(E_{oy}/E_{ox})$ since we can write

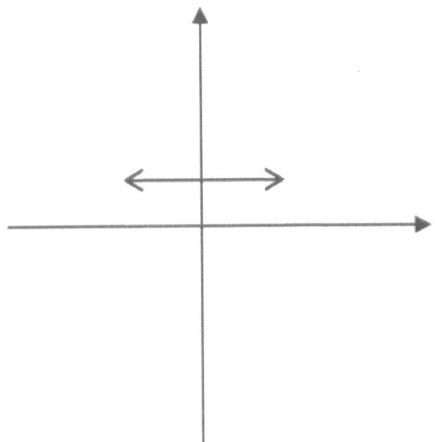

$$\tilde{E}_o = \begin{bmatrix} \tilde{E}_{ox} \\ \tilde{E}_{oy} \end{bmatrix} = A(-1)^m \begin{bmatrix} \cos\alpha \\ \sin\alpha \end{bmatrix} \qquad (5.37)$$

5.8. Jones vector for circular polarization

Suppose $E_{ox} = E_{oy} = A$ and E_x leads E_y by $90° = \pi/2$
At the instant E_x reaches its maximum displacement (+A), E_y is zero. A fourth of a period later, E_x is zero and $E_y = +A$. In order to clarify this, let us

consider the wave at z=0 and choose $\varphi_x=0$ and $\varphi_y=\varepsilon$, so that $\varphi_x > \varphi_y$. Then our E-fields are:

$$E_x = Ae^{-i(\omega t)}$$
$$E_y = Ae^{-i(\omega t - \varepsilon)} \quad (5.38)$$

The negative sign before ε indicates a lag in the y-vibration, relative to x-vibration

$$E_x = A\cos\omega t$$
$$E_y = A\cos\left(\omega t - \frac{\pi}{2}\right) = A\sin\omega t \quad (5.39)$$

Knowing that the angular frequency $\omega=2\pi/T$, the path travelled by the E-vector is easily derived. Also, since $E^2 = E_x^2 + E_y^2 = A^2(\cos^2\omega t + \sin^2\omega t) = A^2$, the tip of the arrow traces out a circle of radius A.

For these cases it is necessary to make $\varphi_y > \varphi_x$ because we have chosen our phase such that the time dependent term (ωt) is negative

$$E_x = E_{ox}e^{i(kz-\omega t+\varphi_x)}$$
$$E_y = E_{oy}e^{i(kz-\omega t+\varphi_y)} \quad (5.40)$$

In order to clarify this, consider the wave at z=0. Choose $\varphi_x=0$ and $\varphi_y=\varepsilon$, so that $\varphi_x > \varphi_y$. Then our E-fields are

$$E_x = Ae^{-i(\omega t)}$$
$$E_y = Ae^{-i(\omega t - \varepsilon)}$$
(5.41)

The negative sign before ε indicates a lag in the y-vibration, relative to x-vibration

5.9. Jones matrix for a rotator

An E-vector oscillating linearly at θ is rotated by an angle, β. Light must be converted to one that oscillates linearly at $(\beta + \theta)$. We then have:

$$\begin{bmatrix} a & b \\ c & d \end{bmatrix} \begin{bmatrix} \cos\theta \\ \sin\theta \end{bmatrix} = \begin{bmatrix} \cos(\beta+\theta) \\ \sin(\beta+\theta) \end{bmatrix} \qquad (5.42)$$

$$M = \begin{bmatrix} \cos\beta & -\sin\beta \\ \sin\beta & \cos\beta \end{bmatrix} \qquad (5.43)$$

Problems

5.1. Describe the state of polarization of the following waves:

$$\vec{E} = E_0[\vec{i}\cos(kz-\omega t) + \vec{j}\sin(kz-\omega t)]$$

$$\vec{E} = E_0[\vec{i}\cos(kz-\omega t) + \vec{j}\sin(kz-\omega t + \frac{\pi}{4})]$$

$$\vec{E} = E_0\vec{i}\cos(kz-\omega t) + 2\vec{j}E_0\sin(kz-\omega t - \frac{\pi}{6})$$

5.2. Find the intensity polarization state of linearized light at 45° to the horizontal direction after passing through a quarter-wave plate with slow axis horizontal, followed by a half-wave plate with slow axis at 45° to the horizontal direction. Take the initial intensity of the light beam to be one unit.

5.3. A partially polarized light beam is composed of 5W/m2 of polarized light and 2W/m2 of natural light.

a. What is the value of the total irradiance?

b. Calculate the degree of polarization of the beam.

5.4. An incident natural light beam strikes the interface of an air–glass ($n_{ti} = 1.54$) at the angle of $60°$ such that $R_{//} = 0$ and $R_{\perp} = 0.165$. Find the degree of polarization of the transmitted wave.

5.5. Find the orientation and ellipticity of the polarization ellipse for each of the following Jones vector:

$$\begin{bmatrix} 2 \\ 5i \end{bmatrix}; \begin{bmatrix} 3 \\ 1+2i \end{bmatrix}; \begin{bmatrix} 2 \\ 4-5i \end{bmatrix}; \begin{bmatrix} 2+3i \\ 4 \end{bmatrix}; \begin{bmatrix} 5i \\ 2 \end{bmatrix}$$

References

Optics. Principles and Applications, K. K. Sharma. Academic Press, Elsevier, 2006.

Schaum's Outline of Optics, Eugene Hecht, McGraw Hills, 1975

Chapter 6

Review of geometric optics

6.1. Introduction

6.2. The thin lens model

6.3. Laws of geometrical optics

6.3.1. Law of Reflection

6.3.2. The Law refraction

6.4. The law of refraction and Fermat's principle

6.5. Light interaction with a prism

6.1. Introduction

Geometric optics uses light ray to describe image formation by spherical surfaces, lenses, mirrors, and other optical instrument.

Consider a positive thin lens, a real mage of a real object. We can develop the geometrical construction of the image by using only two rays:

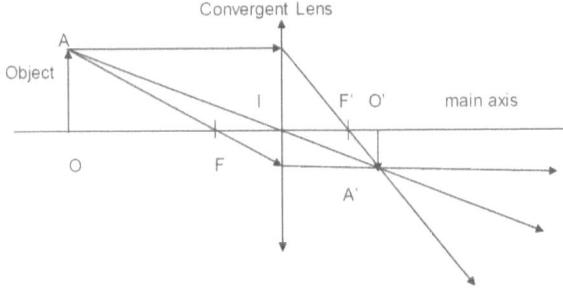

$OI = -x_0$ $O'I = x_i$ and f have the dimension of length, either in centimeter (cm) or meter (m) which is the SI unit. OA represent the object and O'A' the image of the object through the lens

6.2. The thin lens model

With x_0 the object position, x_i the image position and f the focal length, the thin lens model is governed by the equation:

$$\frac{1}{-x_0} + \frac{1}{x_i} = \frac{1}{f} \qquad (6.1)$$

This formula helps to calculate one unknown when parameter the two others are known.

We assume that the line from the object to image point makes only small angles with the axis of the system. This approximation is called the paraxial theory.

Formulas of this type exist for thin lens, thick lens, spherical mirrors, etc…

6.3. Laws of geometrical optics

The ray model states that light propagates as a ray in straight lines with a speed given by:

$c = 2.99792458 \times 10^8 \, m/s$

$c \approx 3 \times 10^3 \, m/s$

The index of refraction of a given medium is $n = \dfrac{c}{v}$ where v, is the speed of light in the medium in question and c, the speed of light in free space.

6.3.1. Law of Reflection

When light in a medium strikes the smooth surface of an object, it will be reflected back in the same medium. This phenomenon is called the incidence phenomenon.

The line (NI) perpendicular to the surface of separation also called interface is called the Normal to the surface of separation.

Light ray represented by the line SI is called the Incident Ray.

Light ray represented by the line IR is called the Reflected Ray.

The point of impact of the incident ray on the surface of separation is denoted by I is called the point of incidence.

Statement of the Law of reflection

When light is reflected by a surface in the same medium, the angle of incidence is equal to the angle of reflection.

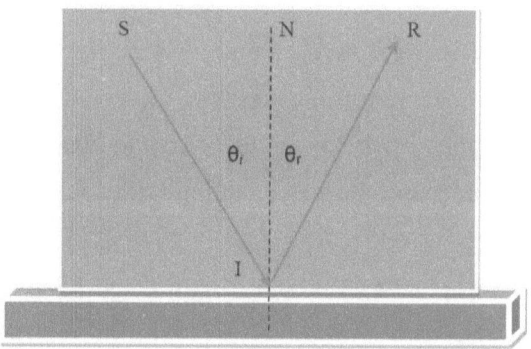

The mathematical expression of the Law of reflection is given by:

$$\theta_i = \theta_r. \qquad (6.2)$$

6.3.2. The Law refraction

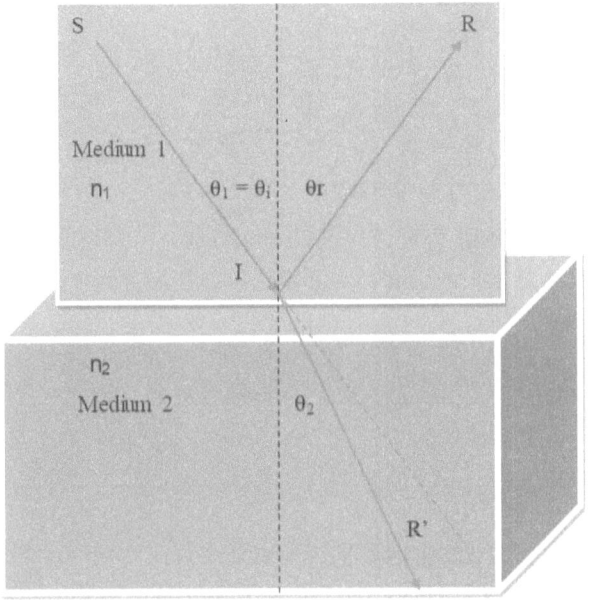

$$n_1 \sin \theta_1 = n_2 \sin \theta_2 \qquad (6.3)$$

IR' is called the refracted ray

The law of reflection is the limiting case of the law of refraction. Indeed, the law of refraction reduces to the law of reflection when $n_1 = n_2$

These two laws could be derived from Maxwell's theory of electromagnetism or from Fermat's principal

6.4. The law of refraction and Fermat's principle
Consider light travelling from a medium 1 to a medium 2 according to the following figure:

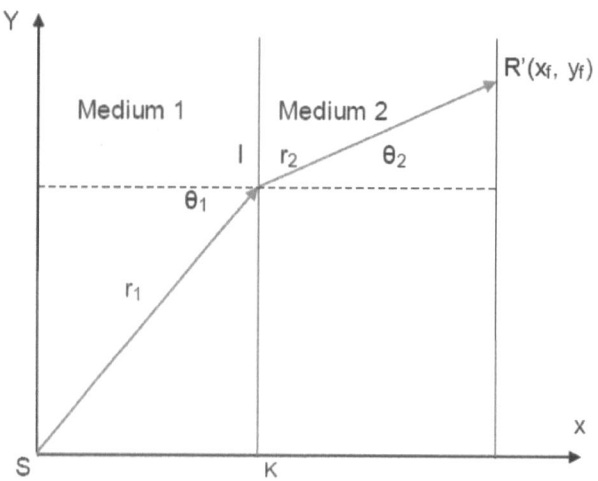

The impact point of the incident light is $I = (x, y)$; $K = (x, 0)$ is the intercept of the surface of separation and the x-axis; $R' = (x_f, y_f)$ is the impact point of the light in medium 2

The path of the light is SIR' and the length of the path is (SR') = SI + IR'= $r_1(y)$ + $r_2(y)$. The optical path is defined as (SR') = $n_1 r_1(y) + n_2 r_2(y)$

The time that light takes to propagate from S to I is

$$t_1 = \frac{r_1(y)}{v_1}$$

The time light takes to travel from I to R' is

$$t_2 = \frac{r_2(y)}{v_2}$$

The total time light takes to travel from S to R' is given by: $t = \dfrac{r_1(y)}{v_1} + \dfrac{r_2(y)}{v_2}$.

By using Pythagoras theorem, $r_1(y) = \sqrt{x^2 + y^2}$ and $r_2(y) = \sqrt{(x_f - x)^2 + (y_f - y)^2}$.

The total time the path of light from S to R' can be then written:

$$t = \frac{\sqrt{x^2 + y^2}}{v_1} + \frac{\sqrt{(x_f - x)^2 + (y_f - y)^2}}{v_2}.$$

The optimum condition requires that the first derivative of the total time with respect to the specific variable y be zero: $\dfrac{dt}{dy} = 0$. In other words,

$$\dfrac{2y}{2v_1\sqrt{x^2+y^2}} + \dfrac{-2(y_f - y)}{2v_2\sqrt{(x_f - x)^2 + (y_f - y)^2}} = 0$$

Or

$$\dfrac{y}{v_1 r_1} - \dfrac{(y_f - y)}{v_2 r_2} = 0$$

But we know that

$$\sin \theta_1 = \dfrac{y}{r_1} = 0 \text{ and } \sin \theta_2 = \dfrac{(y_f - y)}{r_2}$$

As a result, we have $\dfrac{\sin \theta_1}{v_1} = \dfrac{\sin \theta_2}{v_2}$

Using the relation $v = \dfrac{c}{n}$ for each medium, we have the law of Refraction:

$$n_1 \sin \theta_1 = n_2 \sin \theta_2$$

Pierre de Fermat is well known for his work in optics, mathematics

6.5. Light interaction with a prism

A prism scatters light in all direction. This phenomenon called dispersion of light split light in several radiations having each specific wavelength. Consider a prism whose apex angle is A and index of refraction n. An incident ray making an angle of reflection θ_1 with respect to the normal to the prism strikes the prism at the point of incidence. The refracted of an angle θ_2 with respect to the normal to the prism gets refracted again in air of an index of refraction 1 with an angle θ_3 with respect to the normal to the prism. The last ray seems to be incident to the prism with an angle, θ_4. The deviation δ of the final emerging ray with respect to the incident ray is given by:

$$\delta = \theta_1 - \theta_2 + \theta_4 - \theta_3 \tag{6.4}$$

and the apex by $A = \theta_2 + \theta_3$.

With the law of refraction we have $\sin(\theta_1) = n\sin(\theta_2)$ and $n\sin(\theta_3) = \sin(\theta_4)$

We can write the deviation in terms of the apex A:

$$\delta = \theta_1 + \theta_4 - A$$

Let us now express δ as a function of θ_1. From $n\sin(\theta_3) = \sin(\theta_4)$ we have

$$\begin{aligned}
\theta_4 &= Arc\sin\{[n\sin(\theta_3)]\} = Arc\sin\{n\sin(A - \theta_2)\} \\
&= Arc\sin\{n[\sin(A)\sqrt{1 - \sin^2(\theta_2)} - \cos(A)\sin(\theta_2)]\} \\
&= Arc\sin\{n[\sin(A)\sqrt{1 - \frac{\sin^2(\theta_1)}{n^2}} - \cos(A)\frac{\sin(\theta_1)}{n}]\} \\
&= Arc\sin\{[\sin(A)\sqrt{n^2 - \sin^2(\theta_1)} - \cos(A)\sin(\theta_1)]\}
\end{aligned}$$

Therefore,

$$\delta = \theta_1 + Arc\sin\{[\sin(A)\sqrt{n^2 - \sin^2(\theta_1)} - \cos(A)\sin(\theta_1)]\} - A \quad (6.5)$$

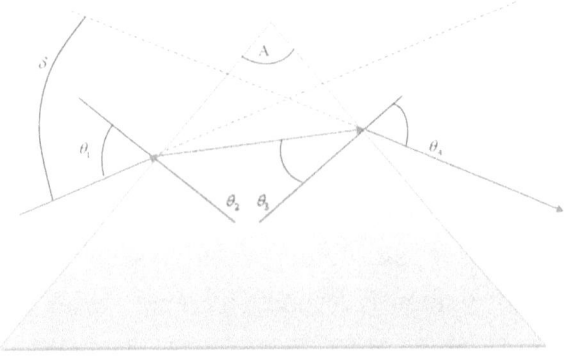

Also, considering that the prism is in air, the incident ray and the emerging ray are both in air of index of refraction approximately 1; we can then write:

$\sin\theta_1 = n\sin\theta_2$ and $n\sin\theta_3 = \sin\theta_4$

Eliminating θ_2 and θ_4 using $\theta_2 = A - \theta_3$ and $\theta_4 = \delta - \theta_1 + A$ we get the two following system of equations:

$$\begin{cases} \sin\theta_1 = n\sin(A-\theta_3) \\ n\sin\theta_3 = \sin(\delta-\theta_1+A) \end{cases}$$
(6.6)

By now differentiating both equations with respect to θ_1 and θ_3, we get the following;

$$\begin{cases} \cos(\theta_1)d\theta_1 + n\cos(A-\theta_3)d\theta_3 = 0 \\ -\cos(\delta-\theta_1+A)d\theta_1 - n\cos(\theta_3)d\theta_3 = 0 \end{cases}$$
(6.7)

We now have an homogenous system of two equations without second member with two unknowns $d\theta_1$ and $d\theta_3$. In order for the system to have a non trivial solution that is a solution which is not zero (0, 0), the determinant of the system has to equal zero. In other words,

$$\begin{vmatrix} \cos(\theta_1) & n\cos(A-\theta_3) \\ -\cos(\delta-\theta_1+A) & -n\cos(\theta_3) \end{vmatrix} = 0 \quad (6.8)$$

$$\cos(\theta_1)\cos(\theta_3) - \cos(A-\theta_3)\cos(\delta-\theta_1+A) = 0 \quad (6.9)$$

The optimum condition where the minimum deviation occurs correspond to the situation where the angle of incidence θ_1 equals the angle of refraction in air, θ_4. Under this condition the emerging light and the incident light become symmetric. The minimum deviation becomes:

$$\delta_{min} = 2 Arc\sin[n\sin(A/2)] - A \qquad (6.10)$$

Chapter 7

Light interaction with spherical surfaces and related equations

7.1. Light and the convex spherical surfaces

7.2. Object and Image distance: Analysis at the infinity

7.2.1. Object at the infinity

7.2.2. Object between the focal length image F' and the optical center O

7.2.3. Magnification

7.3. Concave spherical surface

7.3.1. Thin lens equation

7.3.2. Spherical mirror

7.1. Light and the convex spherical surfaces

Consider light striking a convex spherical surface as shown in the following figure:

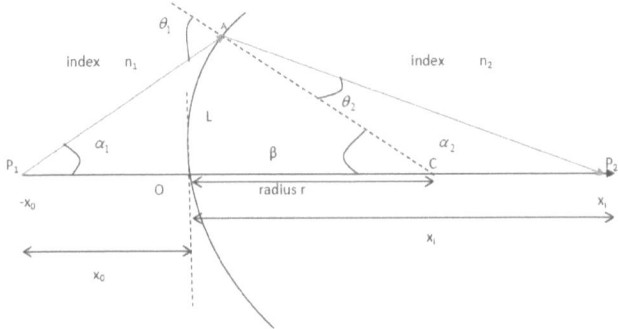

Figure 7.1. Coordinates for the derivation of the paraxial image equation

The convex spherical surface separates medium 1 of index of refraction n_1 from medium 2 of index of refraction n_2. From Snell's Law we can write:

$$n_1 \sin(\theta_1) = n_2 \sin(\theta_2) \tag{7.1}$$

If we use the hypothesis that all the angles are very small so that $\sin(\theta_1) \approx \theta_1$ and $\sin(\theta_2) \approx \theta_2$ then we have:

$$n_1 \theta_1 = n_2 \theta_2. \qquad (7.2)$$

Also, using $\alpha_1 + \beta = \theta_1$ and $\alpha_2 + \theta_2 = \beta$ we can have:

$$\frac{\theta_1}{\theta_2} = \frac{n_2}{n_1} = \frac{\alpha_1 + \beta}{\beta - \alpha_2}$$
(7.3)

The previous relation yields the following:

$$n_1 \alpha_1 + n_2 \alpha_2 = (n_2 - n_1)\beta \qquad (7.4)$$

Now, it is important to mention that the angles being very small, the arc OA is approximately very close to a segment OA that is a straight line of length L. the tangent of these angles are equivalent to the angles in radians. As a result:

$$\tan\alpha_1 = \frac{L}{x_0}, \quad \tan\alpha_2 = \frac{L}{x_i}, \quad \tan\beta = \frac{L}{r} \qquad (7.5)$$

where r is the radius of curvature of the convex spherical surface.
Then we have:

$$\frac{n_1}{x_0} + \frac{n_2}{x_i} = \frac{n_2 - n_1}{r} \qquad (7.6)$$

The previous relation represents the image-forming equation of the convex Spherical surface of radius of curvature r.

Let us now define $\xi_0 = \frac{x_0}{n_1}$, $\xi_i = \frac{x_i}{n_2}$ and $\rho = \frac{r}{n_2 - n_1}$, then the image-forming equation becomes:

$$-\frac{1}{\xi_0} + \frac{1}{\xi_i} = \frac{1}{\rho} \qquad (7.7)$$

Sign convention

We consider point O as the origin of the optical axis ox. Therefore the radius of curvature r is positive for a convex spherical surface and negative for a concave spherical surface

Real and virtual object

When the object is located on the left of the spherical surface it is called real object. The object is called virtual object when it is placed on the right of the spherical surface. A virtual image can be a virtual object to another spherical system. This situation occurs for example for the association of several spherical surfaces. The image given by the first becomes the object of the second and so forth.

7.2. Object and Image distance: Analysis at the infinity

7.2.1. Object at the infinity

The relation between the object and image position with respect to the spherical surface was previously given by the equation $\frac{n_1}{x_0} + \frac{n_2}{x_i} = \frac{n_2 - n_1}{r}$. So for an object placed at the infinity, that is $x_0 \rightarrow +\infty$, the first term of the first member of the relationship vanishes. The equation becomes: $\dfrac{n_2}{x_i} = \dfrac{n_2 - n_1}{r}$ and the image position is given by $x_i = \dfrac{n_2}{n_2 - n_1} r$. Also if we set $\gamma = \dfrac{n_2}{n_1}$, then $x_i = \dfrac{\gamma}{\gamma - 1} r$ with $\gamma \neq 1$, or $n_1 \neq n_2$.

Image at the infinity

For an image at the infinity, that is $x_i \to +\infty$, the second term of the first member of the relationship vanishes. The equation becomes: $\dfrac{n_1}{x_0} = \dfrac{n_2 - n_1}{r}$ and the object position is given by $x_0 = \dfrac{n_1}{n_2 - n_1} r$. Also if we set $\gamma = \dfrac{n_2}{n_1}$, then $x_0 = \dfrac{r}{\gamma - 1}$ with $\gamma \neq 1$, or $n_1 \neq n_2$.

Object and Image construction

Object and Image to the left: object distance greater than the focal length.

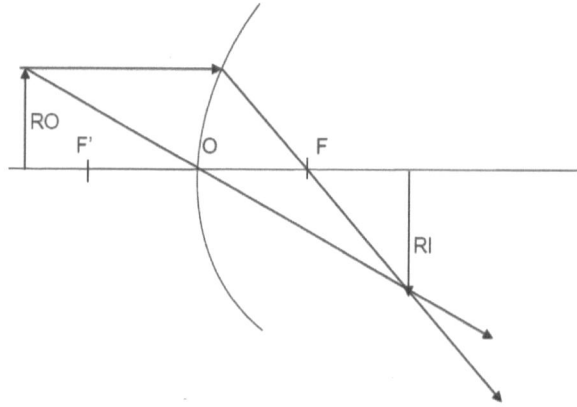

Figure 7.2. Image trough a convex spherical lens
RO = y_o = Real object; RI = y'_o = Real image

The object is represented by an arrow parallel to the y-axis and in the positive direction. An object consists of an infinite number of points. A conjugate point at the image corresponds to each object point. Finally a cone light emerges from each object point and converges to the corresponding conjugate image point. The image also is represented by an arrow parallel to the y-axis but in the negative direction.

7.2.2. Object between the focal length image F' and the optical center O

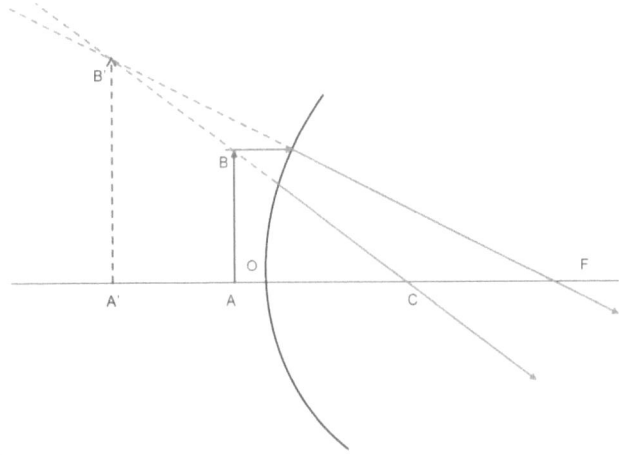

Figure 7.3. Image of an object between the focal length image F' and the optical center O

The previous object AB is real and is represented by a solid line. The corresponding image A'B' is virtual and therefore represented by dashed lines. The object is located between the focal point-image F' and the optical center O. To find the mage we have to trace BC and B'F in dash line. Then they converge at the conjugate image point B' of B. The size of the object is AB = y_o and the size of the image is A'B' = y_i. The image A'B' represented in dashed lines is called virtual image and is completely behind the object. In summary, it is important to remember these three rules when constructing the rays for object and image constructing:

i. Any incident ray parallel to the optical axis emerges from the spherical surface and passes through the focal point object F.

ii. Any incident ray from the object that passes through the center of curvature C of the spherical surface will not deviate.

iii. Any incident ray that passes through the focal point image at will emerge from the spherical surface by being parallel to the optical axis.

Magnification

In all our construction the size of the object is represented by y_o and the size of the image is y_i. The lateral magnification is defined by:

$$m = \frac{y_i}{y_o}. \tag{7.8}$$

Using Thales theorem or the similarity of the right triangle CAB and CA'B', we have:

$$m = \frac{y_i}{y_o} = \frac{x_i - r}{x_o - r} \qquad (7.9)$$

where x_i is the abscise of A' and x_o the one for A.

7.3. Concave spherical surface

We assume in this case that the index of refraction of medium 1 (left) n_1 is less than the index of refraction of medium 2 (right). The image construction follows the same rule as previously used for the construction of image of an object through a convex spherical surface. The image-forming equation is almost the same with the difference that the radius of curvature r is changed to $-$ r for a concave spherical surface. We can give an example of construction:

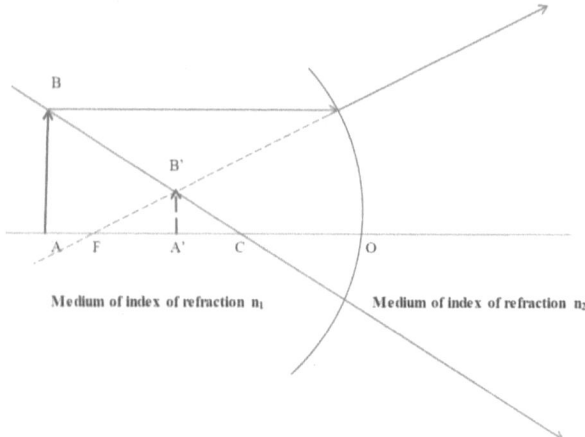

Figure 7.4. Case of the object is located before F

7.3.1. Thin lens equation

A thin lens has two spherical surfaces separated with a very small distance that can be neglected in the derivation of the imaging equation. We can then consider the thin lens as a combination of to spherical surfaces, one convex and the other concave to an observer placed either side of the thin lens.

Derivation of the thin lens equation

Assume n_1 the index of refraction of the medium before the thin lens, n_2 the index of refraction of the medium inside and n_3 the one of the medium after the thin lens. We can have the following picture:

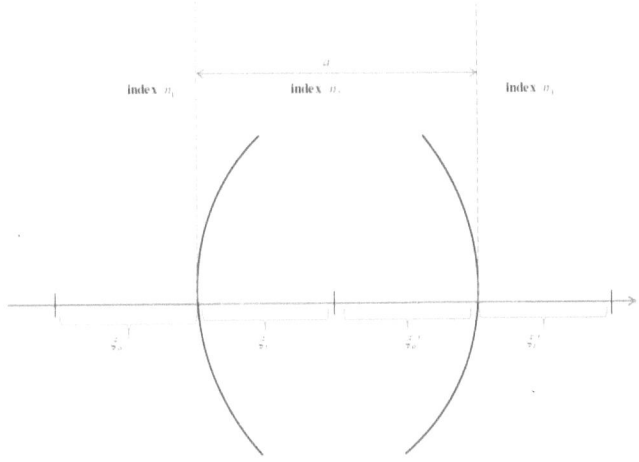

Figure 7.5. Description of a thin lens

The distance between the two optical centers of the two spherical surfaces is a in the medium of index of refraction, n_2. The imaging equation of the first spherical surface is $\dfrac{1}{-\xi_o} + \dfrac{1}{\xi_i} = \dfrac{1}{\rho_1}$

where $\rho_1 = \dfrac{r_1}{n_2 - n_1}$

The imaging equation of the first spherical surface is $\dfrac{1}{-\xi'_o} + \dfrac{1}{\xi'_i} = \dfrac{1}{\rho_2}$

where $\rho_2 = \dfrac{r_2}{n_2 - n_2}$

We know that $\xi = \dfrac{dis\tan ce}{index\ of\ refraction}$; for example $\xi_o = \dfrac{x_o}{n_1}$ and $\xi_i = \dfrac{x_i}{n_2}$. So we do the same for a

$-\xi'_o + \xi_i = \dfrac{a}{n_2}$. ξ_o, ξ_i, ξ'_o, ξ'_i are coordinates of points on the optical axis.

We now obtain the following equations:

$$\begin{cases} \dfrac{1}{-\xi_o} + \dfrac{1}{\xi_i} = \dfrac{1}{\rho_1} \\ -\dfrac{1}{\xi_i - \dfrac{a}{n_2}} + \dfrac{1}{\xi'_i} = \dfrac{1}{\rho_2} \end{cases} \qquad (7.10)$$

For the thickness, $a \to 0$, we have:

$$\begin{cases} \dfrac{1}{-\xi_o} + \dfrac{1}{\xi_i} = \dfrac{1}{\rho_1} \\ -\dfrac{1}{\xi_i} + \dfrac{1}{\xi'_i} = \dfrac{1}{\rho_2} \end{cases} \qquad (7.11)$$

By adding both equations we get

$$\dfrac{1}{-\xi_o} + \dfrac{1}{\xi'_i} = \dfrac{1}{\rho_1} + \dfrac{1}{\rho_2} \qquad (7.12)$$

But $\xi_o = \dfrac{x_o}{n_1}$, $\xi'_i = \dfrac{x'_i}{n_3} = \dfrac{x_i}{n_3}$ because we can now set $x'_i = x_i$; Also

$$-\dfrac{n_1}{x_o} + \dfrac{n_3}{x_i} = \dfrac{1}{\rho_1} + \dfrac{1}{\rho_2} \qquad (7.13)$$

$$\dfrac{1}{\rho_1} + \dfrac{1}{\rho_2} = \dfrac{n_2 - n_1}{r_1} + \dfrac{n_3 - n_2}{r_2} \qquad (7.14)$$

The focal length of the thin lens is defined as

$$\dfrac{1}{f} = \dfrac{n_2 - n_1}{r_1} + \dfrac{n_3 - n_2}{r_2} \qquad (7.15)$$

It is interested to observe the type of equation we would obtain by assuming that $n_1 = n_3 = 1$ represent the index of refraction of air, and $n_2 = n$ the index of refraction of the lens itself. The previous equation becomes:

$$\frac{1}{f} = \frac{n-1}{r_1} + \frac{1-n}{r_2} \qquad (7.16)$$

For a symmetric lens in air $r_1 = -r_2 = r$
This implies that

$$\frac{1}{f} = \frac{n-1}{r} + \frac{1-n}{-r} = \frac{2(n-1)}{r} \qquad (7.17)$$

Also equation $-\frac{n_1}{x_o} + \frac{n_3}{x_i} = \frac{1}{f}$ becomes $-\frac{1}{x_o} + \frac{1}{x_i} = \frac{1}{f}$. We finally obtain a usual lens image-forming equation for a biconvex lens for example. If $f > 0$, we have a positive lens, if $f > 0$, we have a negative lens.

7.3.2. Spherical mirror

Parallel rays striking the concave surface of a spherical concave mirrors get reflected by passing through the center of curvature.

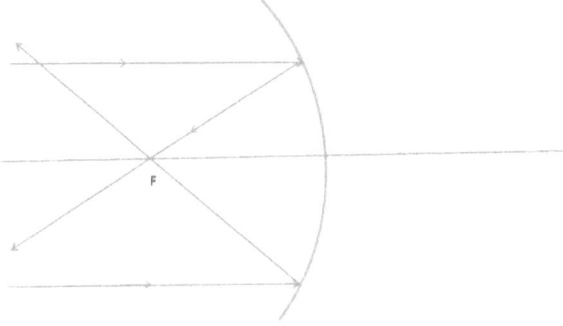

Figure 7.6 Parallel rays emerging though the focal point F

However not all parallel rays converge to F when the curvature of the mirror is too large. This is called spherical aberration.

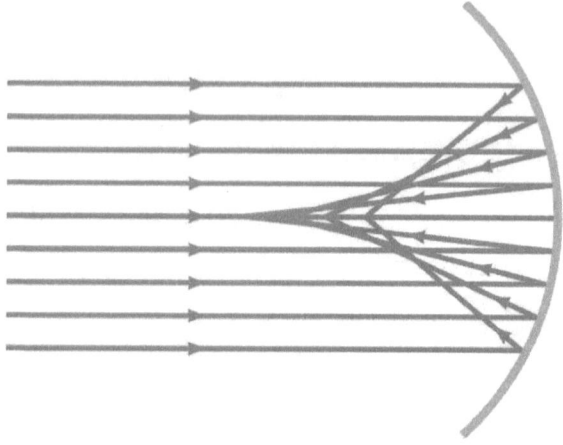

Figure 7.7. Spherical Aberration

Problems

7.1. Prove Snell's by using Pierre de Fermat's Principle

7.2. Describe the story behind Snell's law by writing about Snell, Descartes and Fermat

Chapter 8

Light Interference

8.1. Introduction

8.2. Harmonic waves

8.3 Superposition of the harmonic waves

8.4. Interferences intensities

8.5. Young experiment

8.6. Newton's rings

8.7. The Michelson interferometer

8.8. Fresnel's formulas

8.9. Blackbody radiation

8.1. Introduction

In this chapter, we describe the wave nature of light. In his experiment, Thomas Young observes on a screen an interference pattern that is sequences of bright stripes and not so bright stripes of light.

The interpretation of these patterns was done by using the interference of water waves. But water wave's patterns are made of amplitude interference whereas the super position of light waves is observed as Intensity pattern.

Newton associated light beams with stream of particles and some scientists attached Young thinking that he was diminishing newton work. They are all right (Newton and Young) because it is well known today that light is an electro-magnetic wave and also light has a particle like behavior (quantum mechanics).

In this chapter we will consider the harmonic incident wave, solution of the scalar wave equation, in Cartesian coordinates:

$$\frac{\partial^2 u}{\partial x^2} + \frac{\partial^2 u}{\partial y^2} + \frac{\partial^2 u}{\partial z^2} = \frac{1}{v^2}\frac{\partial^2 u}{\partial t^2} \qquad (8.1)$$

where $v = \dfrac{c}{n}$; n is the index of refraction of the medium.

The previous equation could be written using the operator nabla or $\vec{\nabla}$

$$\vec{\nabla}^2 u + k^2 u = 0 \tag{8.2}$$

where $k = \dfrac{2\pi n}{\lambda}$ is the wave number and λ the wave length of light.

8.2. Harmonic waves

Consider one dimensional wave equation

$$\dfrac{\partial^2 u}{\partial x^2} = \dfrac{1}{v^2}\dfrac{\partial^2 u}{\partial t^2} \tag{8.3}$$

Consider the wave travelling in the vacuum and in an isotropic and non-conducting medium of index of refraction n.

Consider the solution of transversal waves vibrating in the u direction moving in the X-direction and having the wavelength λ and period T

$$u(t) = A\cos\left[2\pi\left(\frac{x}{\lambda} - \frac{t}{T} + \phi\right)\right] \quad (8.4)$$

A is the amplitude of the wave

ϕ is the phase constant

The phase velocity $\vartheta = \frac{\omega}{\kappa}$

The angular frequency $\omega = 2\pi\gamma$

We can write u as

$u = A\cos(kx - wt)$

$= A\cos k\left(x - \frac{\omega}{k}t\right)$

$= A\cos k(x - vt)$

8.3. Superposition of the harmonic waves

Consider the superposition of 2 harmonic waves propagating in the x-direction and vibrating in the y-direction.

$u_1 = A\cos 2\pi\left[\frac{x}{\lambda} - \frac{t}{T}\right]$; $u_2 = A\cos 2\pi\left[\frac{(x-\delta)}{\lambda} - \frac{t}{T}\right]$

The 2 waves have an optical path difference δ.

At time $t = 0$, u_1 has its 1st maximum

At the position $x = 0$ while u_2 has its maximum at $x = \delta$

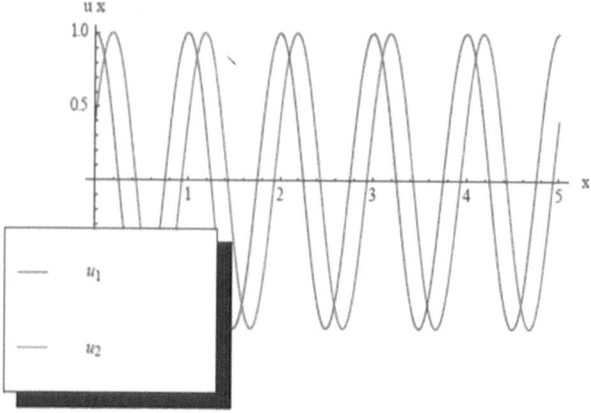

Figure 8.1. Superposition of two wave superposition propagating in the x-direction

The resultant wave $u = u_1 + u_2$

$$u = A\cos 2\pi \left[\frac{x}{\lambda} - \frac{t}{T}\right] + A\cos 2\pi \left[\frac{(x-\delta)}{\lambda} - \frac{t}{T}\right]$$

We will use the following trigonometric properties

$$\cos\alpha + \cos\beta = 2\cos\frac{(\alpha+\beta)}{2}\cos\frac{(\alpha-\beta)}{2}$$

$$u = 2A\cos\left(2\pi\frac{\delta}{2\lambda}\right)\cos 2\pi\left[\left(\frac{x}{\lambda}-\frac{t}{T}\right)-\frac{\delta}{2\lambda}\right]$$

The amplitude of the resultant wave is now

$$\Re = 2A\cos\left(2\pi\frac{\delta}{2\lambda}\right) \qquad (8.5)$$

$$\Re^2 = 4A^2\cos^2\left(\pi\frac{\delta}{\lambda}\right) \qquad (8.6)$$

We can now distinguish 2 cases:

The first case is the one for which the amplitude of the resultant wave u is maximum if $\cos\pi\frac{\delta}{\lambda} = \pm 1$

Or $\pi\frac{\delta}{\lambda} = 0 \pm m\pi$

Or $\delta = m\lambda$ where m is an integer

$\Rightarrow$ Then $u_{max} = 4A^2$

In the second case we consider the amplitude of the resultant wave being zero. In this case,

$$\cos^2\frac{\pi\delta}{\lambda} = 0 \Leftrightarrow \cos\frac{\pi\delta}{\lambda} = 0$$

$$\Leftrightarrow \frac{\pi\delta}{\lambda} = \frac{\pi}{2} + m\pi \quad \Leftrightarrow \delta = \frac{\lambda}{\pi}\left[\frac{\pi}{2} + m\pi\right]$$

$\Leftrightarrow \delta = \lambda\left(m + \frac{1}{2}\right)$. The minimum value of the amplitude is described here by $\Re = 0$

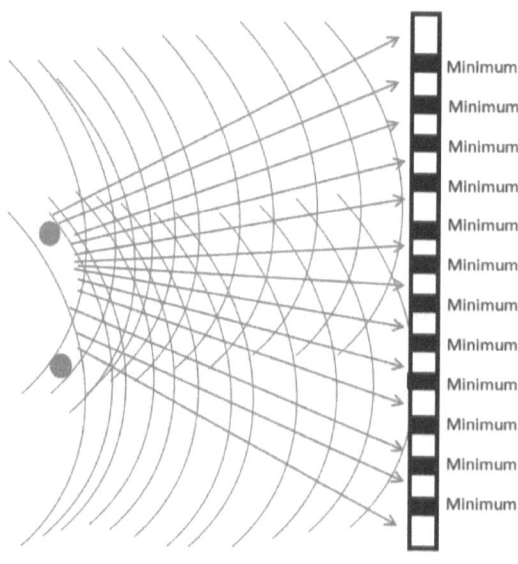

Figure 8.2. Schematic of the interference pattern produced by two waves sources.

8.4. Interferences intensities

The interference pattern of the water is an amplitude pattern while
the interference pattern for light is the intensity pattern, showing
very bright spots I_{max} and dark spots I_{min} in space. If we square
the resultant wave from the superposition we have

$$u^2 = \left[2A\cos\left(2\pi\frac{\delta}{2\lambda}\right)\right]^2 \cos^2\left[\frac{2\pi x}{\lambda} - 2\pi\frac{t}{T} - \frac{2\pi\delta}{2\lambda}\right] \quad (8.7)$$

Let us use of the time average to find I. We will use time average of the cosine square function within a period.

$$\cos^2 t = \frac{1}{T}\int_0^T \cos^2 dt = \frac{1}{2}$$

If we do the same to $\cos^2 2\pi\left[\left(\frac{x}{\lambda} - \frac{t}{T}\right) - \frac{\delta}{2\lambda}\right]$

We will have:

$$a_v^2 = \frac{1}{T}\int_0^T \cos^2 2\pi\left[\frac{x}{\lambda} - \frac{t}{T} - \frac{\delta}{2\lambda}\right]dt = \frac{1}{2} \quad (8.8)$$

We obtain the intensity of the Wave:

$$I = 2A\cos^2\left(\frac{\pi\delta}{\lambda}\right) \cdot \frac{1}{2} \qquad (8.9)$$

8.5. Young experiment

A monochromatic wave is split in two waves by two small openings. On the screen we see several fringes consisting of maxima (bright) and minima (dark) pattern.

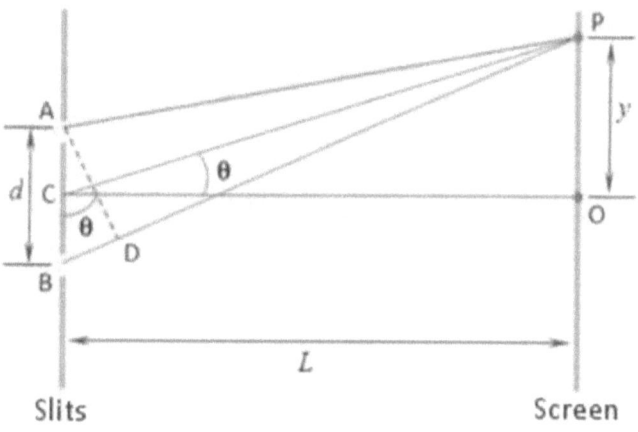

Figure 8.3. Schematic of Young experiment

In our model we will assume that the 2 monochromatic waves leaving the 2 slits $p_1 < p_2$ travel in parallel in the same direction. p_1 stands for B and p_2 for A. They have an angle θ with respect to the symmetry axis. $a = d$ and L=x and the optical path difference $\delta = a\sin\theta$

Proof

$\sin\theta = \dfrac{\delta}{a} \Leftrightarrow \delta = a\sin\theta$ But we consider θ very small.

$\delta = a\sin\theta$

Also $\tan\theta = \dfrac{y}{x} = \sin\theta$

$\Rightarrow \delta = a\dfrac{y}{x}$

$\Rightarrow I(y) = I_0 \cos^2\left[\dfrac{\pi}{\lambda} a \dfrac{y}{x}\right]$

For $y = 0$ $\cos = 1$

$I = I_0 =$ Incident light intensity

Destructive interference is observed for

$$\Rightarrow \cos^2\left[\frac{\pi}{\lambda}a\frac{y}{x}\right] = 0$$

Or $\dfrac{\pi a y}{\lambda x} = \dfrac{\pi}{2} + n\pi$

Or $\dfrac{ay}{x} = \delta = \dfrac{\lambda}{\pi}\left[\dfrac{\pi}{2} + n\pi\right]$

$$= \lambda\left(\frac{1}{2} + n\right)$$

$$= \frac{\lambda}{2},\ 3\frac{\lambda}{2},\ \frac{5}{2}\lambda,$$

Constructive interference is observed for…

8.6. Newton's rings

Consider a convex surface of a lens in contact with a plane glass

Put the design here

There is a thin film of air between the two optical surfaces. When a monochromatic light is incident to the system (see figure), it results the observation of the circular interference fringes as in the following figure:

Put the figure here

One observes alternatively dark and bright rings. These rings studied by Newton are referred to as Newton's Rings.

8.7. The Michelson interferometer

In 1880 the great physicist Albert Michelson used the Michelson interferometer to make precise measurements of wavelengths and very small distances. Michelson Interferometer finally proved that there is no ether as some believed in those days. For Michelson's experiment the source of light produces a monochromatic light. It is a single source. The monochromatic will be split by a beam splitter in two waves with different paths. Interference pattern will be observed when the two light recombine again.

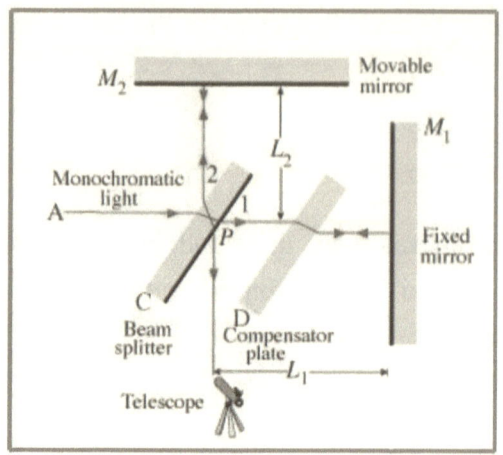

Figure 8.4. Michelson Interferometer experiment setup

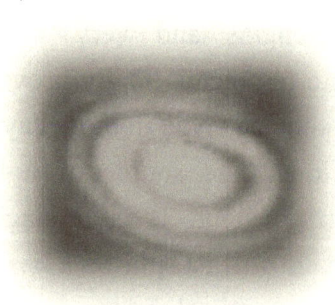

Figure 8.5. Interference pattern produced with a Michelson Interferometer using a green laser

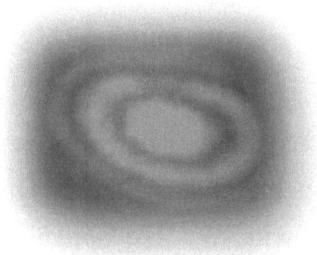

Figure 8.6. Michelson interferometer using red fringes from laser interference

The whole apparatus is mounted on a rigid and stable frame and the position of mirror M_2 can be adjusted accurately. A monochromatic light is sent from the light source A to the beam splitter C. The resulted reflected rays are ray 1 that goes up to the Mirror M_1 and ray 2 that goes to Mirror M_2. The Surface P of beam splitter is painted in silver. Ray 1 reflects off the Mirror M_1, passes through the compensator place D and reflects off the surface P. Ray 2 reflects off mirror M_2 and passes through the beam splitter, C. Finally the two rays combine and reach the observer's eyes on the telescope. If L_1 and L_2 are exactly equal and the two mirrors M_1 and M_2 are also exactly perpendicular, then virtual image of M_1 formed by reflection at the silvered surface of plate C coincides with mirror M_2. If L_1 and L_2 are not exactly equal the image of M_1 is displaced slightly from M_2; and if the mirrors M_1 and M_2 are not exactly perpendicular, the image of M_1 makes a slight angle with M_2. The optical path difference is introduced by the displacement of the mirror M_2. If Mirror M_2 is displaced by $n\dfrac{\delta}{2}$, the optical path difference is $n\delta$ for each wave length component of the incident beam. The center position is $\delta = 0$. At that position we have the superposition of all the wavelengths and the interference is constructive. For these interferences in the direction of the detector, no light travels back to the source. Conversely for destructive interference in the direction of the detector, the light travels backward to the source. The amplitude of the input wave is given by $u_1(x) = A\cos\left(2\pi\dfrac{x_2}{\lambda}\right)$. The amplitude representing the superposition of the two waves emerging from the beam splitter is

$$u_1(\delta) = B\cos\left[2\pi\left(\frac{\delta}{2}\right)\lambda\right] \quad \text{with} \quad B \text{ a real constant.}$$

The light intensity of the superposed wave is

$$I(\delta) = [u(\delta)]^2 = \left\{B\cos\left[2\pi\left(\frac{\delta}{2}\right)\lambda\right]\right\}^2 \qquad (8.10)$$

For $\delta = 0$ at the center, $I(\delta) = B^2 =$ the effective intensity at the detector for $\delta = 0$.

Practice problem

Young experiment. Achromatic fringes

Consider a monochromatic source S of light with a wavelength of $\lambda = 0.55$ μm used for the following Young's experiment:

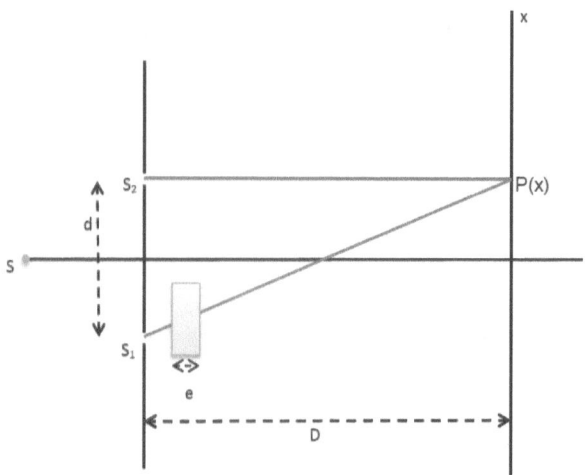

Figure 8.7. Young experiment for achromatic fringes

$S_1S_2 = d$

The slit separation is a=3.3 mm; the distance from the pupil to the screen is D=3m.

Calculate the fringe spacing i

Place a sheet of glass with plane parallel faces and thickness e=0.01mm in front of the slit S_1.

Determine the direction of the displacement of the fringes and the formula giving the relationship for their displacement

Knowing that the fringes are displaced by 4.73 mm, find the index of the glass. How precise is the value of n, if the displacement can be measured to 0.01mm?

Solution

$$i = \lambda \frac{D}{a} = 05\,mm$$

The fringes will displace toward the negative x by the amount: $\Delta x = -\frac{i}{\lambda}(n-1)e$

Index of the glass: $n = 1.5203$

Error determination: $\frac{d(\Delta x)}{x} = \frac{d(\Delta n)}{n-1}$;

$d(\Delta x) = 0.02\,mn$ and $d(n) = 0.5 \times \frac{0.02}{4.73} \approx 0.002$. So finally the index of refraction can be written: $n = 1.5203 \pm 0.002$.

8.8. Fresnel's formulas

Consider light coming from a source and striking a surface separating two media: medium 1 and medium 2 (see figure 6.1)

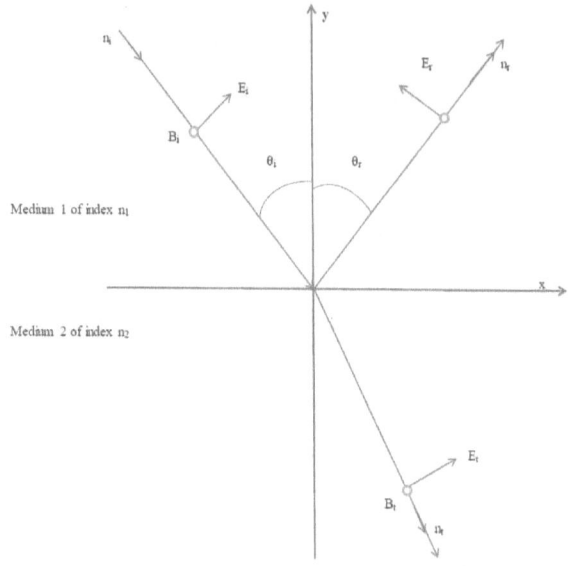

Figure 8.8. Reflection and refraction: the electromagnetic fields **E** and **B**

Because $\vec{E}$ is in the incidence plane, we call it $\vec{E}_{11}$.

$\vec{B}$ is the magnetic field; It is perpendicular to the electric field at each point, and $\vec{n}$ is the normal unit vector

$$\vec{E} = \vec{E}_{011} \exp ik(\vec{n}\vec{r} - \omega t)$$

Because $\vec{B}$ is perpendicular to the incidence plane, we will represent it by $\vec{B}_\perp$

$$\vec{B}_\perp = \vec{B}_{0\perp} \exp ik(\vec{n}\vec{r} - \omega t) \qquad (8.11)$$

The medium is isotropic and nonconductive medium.

$$\vec{n}_i = (\vec{n}_i \vec{i})\vec{i} + (\vec{n}_i j)\vec{j} + (\vec{n}_i \vec{k})\vec{k}$$

$$= \cos\left(\frac{\pi}{2} - \theta_i\right)i - \cos\theta_i \vec{j}$$

$$= \sin\theta_i \vec{i} - \cos\theta_i \vec{j}$$

$$\vec{E}_{1li} = \vec{E}_{011} \exp k(\vec{n}_i \vec{r}_i - \omega t)$$

$$\vec{E}_i = \vec{E}_{io} \exp i\left[k\left(\sin\theta_i \vec{i} - \cos\theta_i \vec{j}\right)\left(X\vec{i} + Y\vec{j}\right) - \omega t\right]$$

Incidence:

$$\vec{E}_i = \vec{E}_{io} \exp i\left[k\left(\sin\theta_i \vec{i} X - \cos\theta_i \vec{j} Y\right) - \omega t\right]$$

$$\vec{n}_r = (\vec{n}_r \vec{i})\vec{i} + (\vec{n}_r \vec{j})\vec{j} + (\vec{n}_r \vec{k})\vec{k}$$

$$\vec{n}_r = \sin\theta_r \vec{i} + \cos\theta_r \vec{j}$$

$$r_r = Xi + Yj$$

$$\vec{E}_r = \vec{E}_{r0} \exp i\left[k\left(\sin\theta_r + \cos\theta_r\right)\left(Xi + Yj\right) - \omega t\right]$$

Reflection:

$$\vec{E}_r = \vec{E}_{r0} \exp i\left[k\left(\sin\theta_r X + \cos\theta_r Y\right) - \omega t\right]$$

$$\bar{n}_t = (\bar{n}_t\bar{i})\bar{i} + (\bar{n}_t\bar{j})\bar{j} + (\bar{n}_r\bar{k})\bar{k}$$
$$\bar{n}_t = \sin\theta_t\bar{i} - \cos\theta_t\bar{i}$$
$$r_t = Xi + Yj$$
$$\bar{E}_t = \bar{E}_{t0}\exp i[k(\sin\theta_t + \cos\theta_t)(Xi + Yj) - \omega t]$$

Transmitted:
$$\bar{E}_t = \bar{E}_{t0}\exp i[k(\sin\theta_t X + \cos\theta_t Y) - \omega t]$$

Boundary condition project on X-axis $\vec{OX}$
$$= \bar{E}_i\cos\theta_i - \bar{E}_r\cos\theta_r = \bar{E}_t\cos\theta_t$$
$$\cos\theta_i\bar{E}_{i0}\exp i[k_1(X\sin\theta_i) - \omega t] - \cos\theta_r E_{r0}\exp i[k_1 X\sin\theta_r - \omega t] = \cos\theta_t\bar{E}_t\exp i[k_2 X\sin\theta_t - \omega t]$$

We first have two write the equations as follows:

(1)...... $k_1\sin\theta_i = k_1\sin\theta_r = k_2\sin\theta_t$.

According to Snell's law

$$\begin{cases} \theta_i = \theta_r \\ n_1\sin\theta_r = n_2\sin\theta_t \end{cases} \quad (8.12)$$

(2)........ $\cos\theta_i E_{i0} - \cos\theta_r E_{r0} = \cos\theta_t E_{t0}$

For the magnetic field

(3)....... $B_i + B_r = B_t$

The relationship between electric and magnetic field is,
$$c = \frac{E}{B} = \frac{v}{n}$$
$$\therefore B = \frac{nE}{v}, v = constant$$

(4)....... $n_1 E_{i0} + n_1 E_{r0} = n_2 E_{t0}$

Reflected amplitude = Reflection coefficient, $r_{11} = \dfrac{E_{r0}}{E_{t0}}$

Transmitted amplitude = Transmission coefficient, $t_{11} = \dfrac{E_{t0}}{E_{i0}}$

Divide equation (2) by E_{i0}

(5)....... $\cos\theta_i - r_{11}\cos\theta_r = t_{11}\cos\theta_t$

Divide equation (4) by E_{i0} and n_1

(6)....... $1 + r_{11} = \dfrac{n_2}{n_1} t_{11}$

Now we can use a matrix form to find the reflectance and transmittance coefficient. But before that let us rearrange equation (5) and (6); we get:

$$\cos\theta_i r_{11} + t_{11}\cos\theta_t = \cos\theta_i$$

$$1 * r_{11} - \frac{n_2}{n_1} t_{11} = -1$$

$$\begin{bmatrix} \cos\theta_r & \cos\theta_t \\ 1 & -n_2/n_1 \end{bmatrix} \begin{bmatrix} r_1 \\ t_{11} \end{bmatrix} = \begin{bmatrix} \cos\theta_i \\ -1 \end{bmatrix}$$

$$\begin{bmatrix} r_1 \\ t_{11} \end{bmatrix} = \begin{bmatrix} \cos\theta_r & \cos\theta_t \\ 1 & -n_2/n_1 \end{bmatrix} \begin{bmatrix} \cos\theta_i \\ -1 \end{bmatrix}$$

If $A = \begin{bmatrix} \cos\theta_r & \cos\theta_t \\ 1 & -n_2/n_1 \end{bmatrix}$

$$|A| = \det(A) = -\frac{n_2}{n_1}\cos\theta_r - \cos\theta_t$$

$\theta_i \neq \pi/2, \theta_r \neq \pi/2$ and $|A| \neq 0$

$$A^{-1} = \frac{1}{|A|} \begin{bmatrix} -n_2/n_1 & -\cos\theta_t \\ -1 & \cos\theta_r \end{bmatrix}$$

$$\begin{bmatrix} r_1 \\ t_{11} \end{bmatrix} = \begin{bmatrix} -\dfrac{n_2}{n_1|A|} & \dfrac{-\cos\theta_t}{|A|} \\ \dfrac{-1}{|A|} & \dfrac{\cos\theta_r}{|A|} \end{bmatrix} \begin{bmatrix} \cos\theta_i \\ -1 \end{bmatrix}$$

$$r_{11} = -\frac{n_2}{n_1}\cos\theta_1 + \frac{\cos\theta_t}{|A|} = \frac{-n_2\cos\theta_i + n_1\cos\theta_t}{-n_2\cos\theta_r - n_1\cos\theta_t}$$

$$t_{11} = \frac{-\cos\theta_i}{|A|} - \frac{\cos\theta_r}{|A|} = \frac{-n_1\cos\theta_i - n_1\cos\theta_r}{-n_2\cos\theta_r, n_1\cos\theta_t}$$

Because $\theta_i = \theta_r$ in Snell's law.

$$r_{11} = \frac{n_2\cos\theta_i - n_1\cos\theta_t}{n_2\cos\theta_i + n_1\cos\theta_t}$$

(8.13)

$$t_{11} = \frac{2n_1\cos\theta_i}{n_2\cos\theta_i + n_1\cos\theta_t}$$

For Perpendicular, $r_\perp$ and $t_\perp$

$$r_\perp = \left(\frac{E_{r0}}{E_{i0}}\right)_\perp , \quad t_\perp = \left(\frac{E_{t0}}{E_{i0}}\right)_\perp \qquad (8.14)$$

Similarly we can write,

$$r_\perp = \frac{n_1 \cos\theta_i - n_2 \cos\theta_t}{n_1 \cos\theta_i + n_2 \cos\theta_t}$$

$$t_\perp = \frac{2n_1 \cos\theta_i}{n_1 \cos\theta_i + n_2 \cos\theta_t} \qquad (8.15)$$

The coefficient of Reflectance, $R = r^2$ and the coefficient Transmittance, $T = \dfrac{n_t \cos\theta_t}{n_i \cos\theta_i} t^2$

$$R + T = r^2 + \left(\frac{n_t \cos\theta_t}{n_i \cos\theta_i}\right) t^2$$

$$= \frac{(n_i \cos\theta_i - n_t \cos\theta_t)^2}{(n_i \cos\theta_i + n_t \cos\theta_t)^2} + \frac{n_t \cos\theta_t}{n_i \cos\theta_i} * \frac{4n_i^2 \cos^2\theta_i}{(n_i \cos\theta_i + n_t \cos\theta_t)^2}$$

$$= \frac{(n_i \cos\theta_i)^2 - 2n_i n_t \cos\theta_i \cos\theta_t + (n_t \cos\theta_t)^2 + 4n_i n_t \cos\theta_i \cos\theta_t}{(n_i \cos\theta_i + n_t \cos\theta_t)^2} \quad (8.16)$$

$$= \frac{(n_i \cos\theta_i)^2 + 2n_i n_t \cos\theta_i \cos\theta_t + (n_t \cos\theta_t)^2}{(n_i \cos\theta_i + n_t \cos\theta_t)^2}$$

$$= 1$$

$R + T = 1$
$T = 1 - R, R = 1 - T$

8.9. Blackbody radiation

A blackbody radiation is the emission of electromagnetic radiation from a closed cavity at a given temperature through a hole.

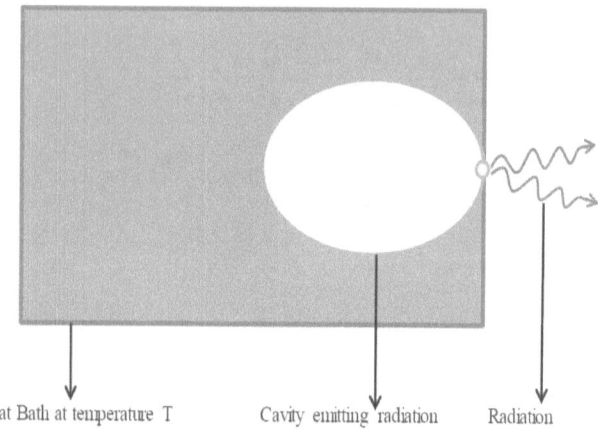

Figure 8.9. A blackbody radiator

It is clear that the cavity completely absorbs the incident radiation and several reflections occur in the container and only a small amount of light get out of it. As a result we can say that the opening is perfectly absorbing or black. The escaped radiation through the hole will be in thermal equilibrium with the matter within the enclosure. The leak of energy is assumed to be very small compare to the total energy in the cavity. The radiation coming out from the black surface is called the blackbody radiation. Planck used a hypothesis of quantized oscillators in a radiating body in 1901 to derive the expression of the hemispherical blackbody spectral radiative flux given by the relation:

$$F_\nu^{bb} = \frac{m_r^2}{c^2} \frac{2\pi h \nu^3}{(e^{h\nu/k_B T} - 1)} \qquad (8.17)$$

With h the Planck constant, m_r, the real index of refraction and k_B

Throughout the cavity the radiation is isotropic and unpolarized. For the net flux to be zero everywhere, we need to have an equal hemispherical flux opposing F_ν^{bb}. In calculations we most of the time would need to compare $h\nu$ and $k_B T$. This will give the Wien's approximation for high energies:

For $h\nu \gg k_B T$

$$F_\nu^{bb} = \frac{m_r^2}{c^2} 2\pi h \nu^3 e^{-h\nu/k_B T} \qquad (8.18)$$

Rayleigh-Jeans limit gives the approximation for low energies where $h\nu \ll k_B T$ and

$$F_v^{bb} = \frac{2\pi k_B T m_r^2 v^2}{c^2} \tag{8.19}$$

Planck function

Planck function is given by:

$$B_v(T) = \frac{F_v^{bb}}{\pi} = \frac{m_r^2}{c^2} \frac{2hv^3}{(e^{hv/k_B T} - 1)} \tag{8.20}$$

which has the unit of an intensity.

The graph of Planck function gives the spectral distribution of the blackbody radiation. For this, the maximum occur for $\lambda_m T = 2,897.8\ \mu m.K$. The maximum temperature gives rise to Wien displacement law which states that *the wavelength of the peak blackbody emission is inversely proportional to temperature.*

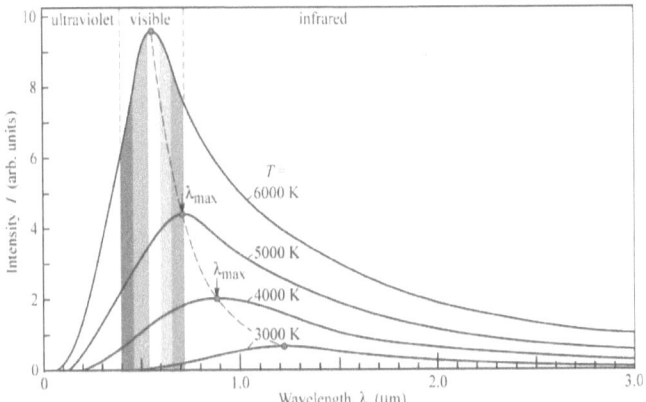

Figure: Blackbody radiation intensity vs. wavelength.

Stefan-Boltzmann law

The frequency-integrated emerging blackbody flux is proportional the fourth power of the absolute temperature:

$$F^{bb} = \sigma_B T^4 \qquad (8.21)$$

The coefficient of proportionality, σ_B is called the Stefan-Boltzmann constant and has the value, $\sigma_B = 5.6703 \times 10^{-8}\ W.m^{-2}.K^{-4}$.

Problems
Derive the Planck function
Derive the Stefan-Boltzmann law

Chapter 9

Linear polarization by anisotropy

9.1. Introduction

9.2. Birefringence

9.1. Introduction

The primary form of polarizers was dichroic crystals like tourmaline. These types of substances are made of crystals with long narrow lattice structures orientated parallel to each other. The interaction between light and the substance depends on the direction of the incident beam as well as on the orientation of the plane of vibration with the crystal axis. This type of material is anisotropic.

9.2. Birefringence

When an incident monochromatic light beam hit an interface, it may happen that two outgoing beam are transmitted. When this occurs, at least one of the media is anisotropic. This phenomenon was first observed in 1669 by Dane Erasmus Bartholimus, who called is double refraction. Media in which double refraction occurs are called birefringent. If the incident parallel beam was naturally unpolarized on an anisotropic material with two planar interfaces, the two transmitted beams are linearly polarized along two mutually orthogonal directions. These two directions are determined by the orientation of the anisotropic material. When the two sides of the anisotropic material are planar and parallel, the two emerging beams are parallel to the incident one. In the case of the figure 9.1 for example, the intensity of the two transmitted light are equal. For figure 9.2, a polarizer is introduced in front of the incident light beam and before the anisotropic material. The relative intensity of the emerging light beam depends on the orientation of the polarizer. In figure 9.3 and 9.4 for two special orientations of the polarizer, only one beam is transmitted. These special orientations are called privilege direction of vibrations. The orientations of the vibrations depends not only on the anisotropic material but also on the direction of propagation

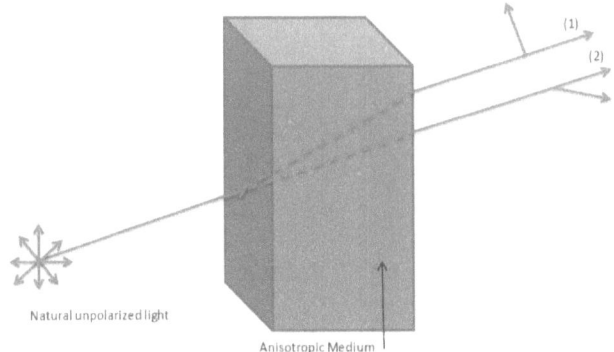

Figure 9.1. Birefringence without a polarizer. The two transmitted beams observed are polarized linearly and orthogonally.

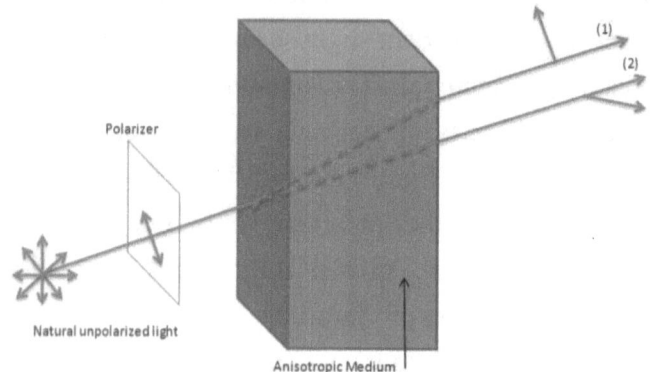

Figure 9.2. Birefringence with polarizer with an arbitrary direction.

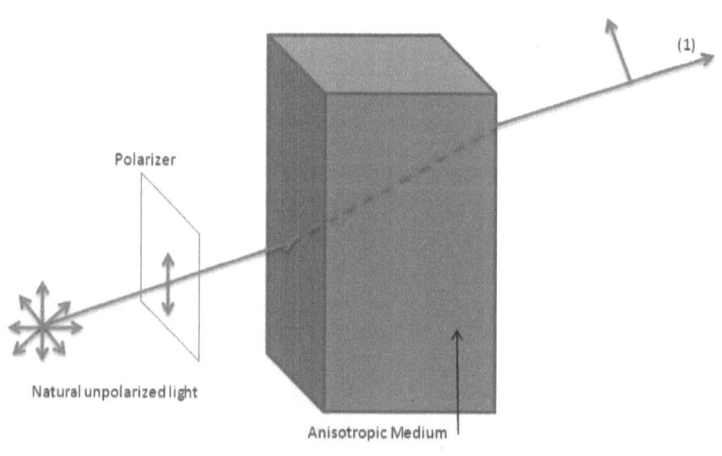

Figure 9.3. birefringence with a polarizer parallel to one of the two privileged vibrations

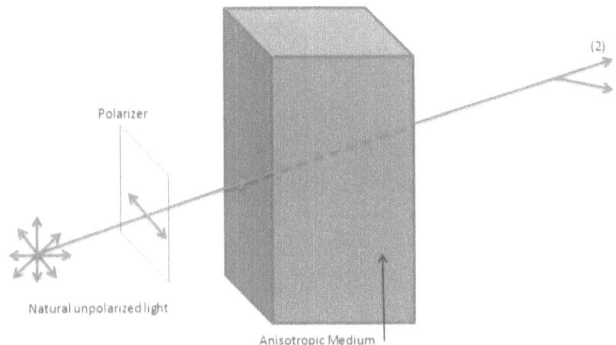

Figure 9.4. birefringence with a polarizer parallel to the other privileged vibration

It is important to important to find the direction in the crystal where the phenomenon does not occur. That direction will be the optical axis. But in the birefringent crystal electrons can move with different amount of freedom in different directions. An example is the Calcite or Iceland spar seen in figure 9.5. which is a crystalline form of the Calcium Carbonate of molecular formula $CaCO_3$. The crystalline has a rhombohedra structure. All the sides are parallelogram of angles 78° and 102°. All the corners except two contain this two angles

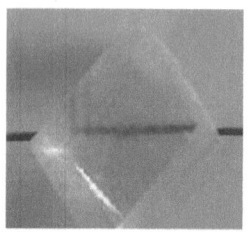 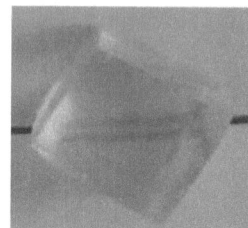

Figure 9.5. Example of natural Calcite spar

Calcite has only one direction with zero effect and therefore belongs to the class of uniaxial crystal. Two types of rays emerge out of the crystal: the ordinary ray (o-ray) and the extraordinary ray (e-ray). The two rays can be seen in figure 9.6 below. The ordinary ray behaves just like if the medium was isotropic while the extraordinary ray takes a different direction within the anisotropic medium. These two rays are linearly polarized even though the incident ray was unpolarized. The velocity of light and the refractive index within the material are constant in all direction for the ordinary ray. These parameters differ in different directions for the extraordinary ray. The two velocities are equal only along the optical axis.

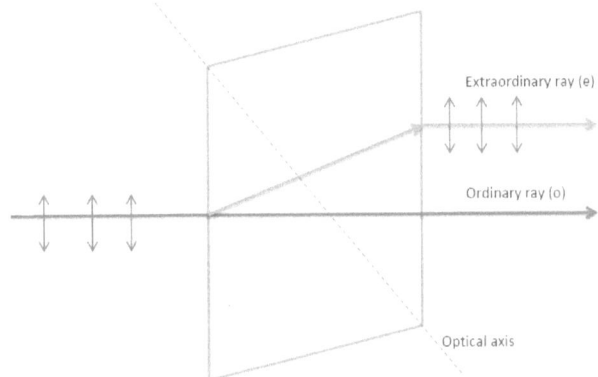

Figure 9.6. Double refraction by calcite of an unpolarized beam incident normally in the section containing the optical axis.

Chapter 10

Fourier transform spectroscopy

10.1. Introduction

10.2. Fourier series

10.3. Fourier transform

10.3.1. Definition of Fourier transform

10.3.2. Properties of the Fourier transform

10.4. Fourier transform spectroscopy

10.1. Introduction

10.2. Fourier series

A quantity that varies periodically with time like waves, rotating machines, may be described by periodic functions. A function is said to be periodic if there exists a real number T > 0 such that for every value t, $f(t+nT) = f(t)$ where n is a positive or negative integer. The smallest value T is called the period of the function f. If t is the time variable, then $f(t)$ is a periodic function of period T. The fundamental angular frequency of the function f is $\omega = \dfrac{2\pi}{T}$. The function $f(t)$ can be written as a Fourier series, that is, into an infinite sum of harmonics components at multiples of the fundamental angular frequency:

$$f(t) = a_0 + \sum_{n=1}^{\infty} [a_n \cos(n\omega t) + b_n \sin(n\omega t)]$$
$$= a_0 + \sum_{n=1}^{\infty} c_n \cos(n\omega t + \phi_n)$$

(10.1)

$$f(t) = a_0 + \sum_{n=1}^{\infty} d_n \sin(n\omega t + \psi_n)$$
$$= \sum_{n=-\infty}^{n=\infty} g_n \exp(in\omega t) \qquad (10.2)$$

The coefficients a_n, b_n, c_n, d_n, g_n are called amplitudes, ϕ_n and ψ_n are called the initial phases or the phase at time t=0. The determination of these coefficients is the most important problem in harmonic analysis. These coefficients are given by the Euler formulas:

$$a_0 = \frac{1}{T} \int_0^T f(t) \, dt \qquad (10.3)$$

$$a_n = \frac{2}{T} \int_0^T f(t) \cos(n\omega t) \, dt \text{ for } n \geq 1 \qquad (10.4)$$

$$b_n = \frac{2}{T} \int_0^T f(t) \sin(n\omega t) \, dt \text{ for } n \geq 1 \qquad (10.5)$$

$$g_n = \frac{1}{T} \int_0^T f(t) e^{in\omega t} \, dt \qquad (10.6)$$

with $n = 0, \pm 1, \pm 2, \pm 3, \pm 4, \ldots$

The function f is assumed to be absolutely integrable. The integration sign goes over a full period and can be taken from t_0 to $t_0 + T$, or from $-\frac{T}{2}$ to $\frac{T}{2}$. Anyway, it has to be chosen in an appropriate way so that the difference between the upper and the lower value of the integration sign is one period, that is T. The following relationships between the coefficients can be proved for $n \geq 1$:

$$a_n = g_n + g_{-n} = c_n \cos\phi_n = d_n \sin\psi_n$$
$$b_n = i(g_n - g_{-n}) = -c_n \sin\phi_n = d_n \cos\psi_n$$
$$g_n = \frac{1}{2}(a_n - ib_n) \text{ and } g_{-n} = \frac{1}{2}(a_n + ib_n)$$
$$c_n^2 = d_n^2 = a_n^2 + b_n^2 = 4 g_n g_{-n} \text{ (sign ambiguity)}$$
$$\tan\phi_n = \frac{-b_n}{a_n} \text{ and } \tan\psi_n = \frac{a_n}{b_n} \text{ (phase ambiguity)}$$

a_0 can be regarded as the mean value of the function f

It is important to note that the function $f(t)$ can be a complex function and t can be stretched from $t = -\infty$ to $t = +\infty$. In such case the Fourier series is called complex Fourier series and the above equation holds

Example using Mathematica Command

For the 3rd order Fourier series of t/2

FourierSeries[t/2, t, 3]

Plot[%, {t, −3Pi, 3Pi}, AxesLabel → {t, FourierSeries}]

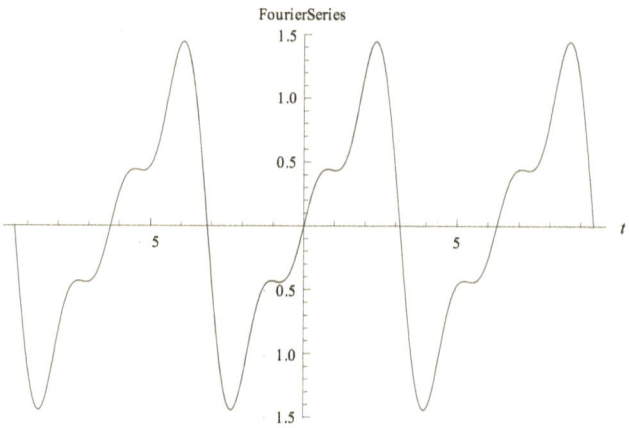

Figure 10.1. Graph of the third order Fourier series of t/2

For the fourth order Fourier series of t

FourierSeries[t, t, 4]

`Plot[%, {t, -4Pi, 4Pi}, AxesLabel → {t, FourierSeries}]`

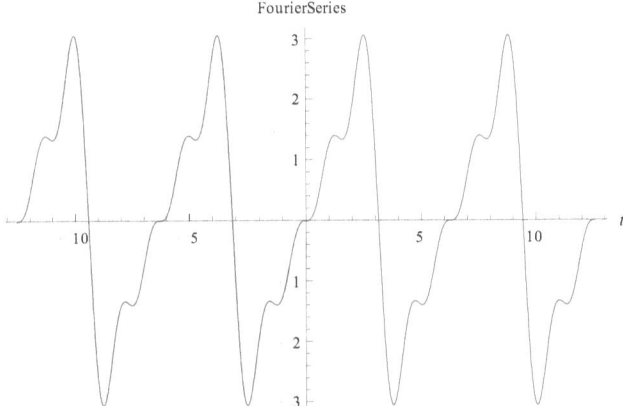

Figure 10.2. Graph of the fourth order Fourier series of t

In other definitions of the Fourier series, in the expansion of the function f a_0 is replace by $\dfrac{a_0}{2}$.

For example if the function f is defined for all x such that $-\pi \leq x \leq \pi$ then it can be represented as the series:

$$f(x) = \frac{a_0}{2} + \sum_{n=1}^{\infty}[a_n \cos(nx) + b_n \sin(nx)] \text{ and }$$

the Fourier coefficients of $f(x)$ can be given by the Euler formulas:

$$a_n = \frac{1}{\pi}\int_{-\pi}^{\pi} f(x)\cos(nx)\,dx \text{ for } n \geq 1 \quad (10.7)$$

$$b_n = \frac{1}{\pi}\int_{-\pi}^{\pi} f(x)\sin(nx)\,dx \text{ for } n \geq 1 \quad (10.8)$$

The coefficient ½ of a_0 is not a big deal and one should not be confused by that since a_0 is always a constant value here. Let's now give a few examples of computations of the Fourier coefficients:

Example 1
Compute the Fourier coefficients of the function $f(x) = x^2 + \pi x$

Solution
$$a_0 = \frac{1}{\pi}\int_{-\pi}^{\pi}(x^2 + \pi x)\,dx = \frac{2}{3}\pi^2$$

For $n \geq 1$,
$$a_n = \frac{4}{n^2}\cos(n\pi)$$
$$b_n = \frac{-2\pi}{n}\cos(n\pi)$$

Example 2

Compute the Fourier coefficients of the function defined as follows:

$$f(x) = \begin{cases} \frac{1}{3}(x+\pi) & \text{if } -\pi < x < \frac{\pi}{2} \\ \pi - x & \text{if } \frac{\pi}{2} < x < \pi \end{cases}$$

Solution

$$a_0 = \frac{1}{\pi}\int_{-\pi}^{\pi} f(x)\,dx = \frac{\pi}{2}$$

For $n \geq 1$,

$$a_n = \frac{2}{3n^2\pi}\left[\cos(n\pi) - \cos\left(\frac{n\pi}{2}\right) + \frac{n\pi}{2}\sin\left(\frac{n\pi}{2}\right)\right]$$

$$b_n = \frac{2}{3n^2\pi}\left[-3n\pi\cos(n\pi) + \frac{3n\pi}{2}\sin\left(\frac{n\pi}{2}\right) - \sin\left(\frac{n\pi}{2}\right)\right]$$

10.3. Fourier transform

While Fourier series allows a periodic function to be written as a series of harmonic oscillations at definite frequencies, Fourier transform allows an aperiodic function to be expressed as an integral sum over a continues range of frequencies. The following integrals are called Fourier integrals:

$$f(t) = \int_0^\infty [A(\omega)\cos(\omega t) + B(\omega)\sin(\omega t)]\, d\omega$$

$$= \int_0^\infty C(\omega)\cos[\omega t + \phi(\omega)]\, d\omega$$

$$= \int_0^\infty D(\omega)\sin[\omega t + \psi(\omega)]\, d\omega$$

$$= \frac{1}{\sqrt{2\pi}} \int_{-\infty}^{+\infty} E(\omega) e^{+i\omega t}\, d\omega \qquad (10.9)$$

$$= \frac{1}{2\pi} \int_{-\infty}^{+\infty} F(\omega) e^{+i\omega t}\, d\omega$$

$$= \int_{-\infty}^{+\infty} G(v) e^{+i2\pi v t}\, d\omega$$

The determination of the functions, $A(\omega)$, $B(\omega)$, $C(\omega)$, $D(\omega)$, $E(\omega)$, $F(\omega)$, $G(v)$ is the central problem of Fourier analysis. Each of the functions $E(\omega)$, $F(\omega)$, $G(v)$ is known as the Fourier transform of the function $f(t)$. Indeed there is none universally accepted convention as to which is meant by title. In all the above equations, ω represents an angular frequency in radians per second while v represents the frequency in cycles per second. The fo
llowing relationships between functions hold:

$$A(\omega) = \frac{1}{\pi} \int_{-\infty}^{+\infty} f(t) \cos(\omega t) dt$$
$$= \frac{1}{2\pi} [F(\omega) + F(-\omega)] \qquad (10.10)$$
$$= C(\omega) \cos[\phi(\omega)] = D(\omega) \sin[\psi(\omega)]$$

$$B(\omega) = \frac{1}{\pi} \int_{-\infty}^{+\infty} f(t) \sin(\omega t) dt$$
$$= \frac{i}{2\pi} [F(\omega) - F(-\omega)] \qquad (10.11)$$
$$= -C(\omega) \sin[\phi(\omega)] = D(\omega) \cos[\psi(\omega)]$$

$$E(\omega) = \frac{1}{\sqrt{2\pi}} \int_{-\infty}^{+\infty} f(t) e^{-i\omega t} dt$$
$$= \frac{1}{\sqrt{2\pi}} G(\omega / 2\pi) \qquad (10.12)$$

$$F(\omega) = \int_{-\infty}^{+\infty} f(t) e^{-i\omega t} dt$$
$$= \pi [A(\omega) - iB(\omega)] \quad \text{for } \omega > 0 \qquad (10.13)$$
$$= \pi [A(|\omega|) + iB(|\omega|)] \quad \text{for } \omega < 0$$

$$G(\nu) = \int_{-\infty}^{+\infty} f(t) e^{-i 2\pi \nu t} dt = F(2\pi \nu) \qquad (10.14)$$

It can also be proved that:

$$C^2(\omega) = D^2(\omega) = A^2(\omega) + B^2(\omega)$$
$$= \frac{1}{\pi^2} F(\omega) F(\omega)$$
$$\tan[\phi(\omega)] = -\frac{B(\omega)}{A(\omega)} \qquad (10.15)$$
$$\tan[\psi(\omega)] = \frac{A(\omega)}{B(\omega)}$$

10.3.1. Definition of Fourier transform

It is understood that there is no universally accepted convention in terms of the definition of the Fourier transform and inverse Fourier transform. If we use the variable x and y instead of t and ω, we can define the Fourier transform and the Fourier inverse transform as follows:

$$F(y) = \int_{-\infty}^{+\infty} f(x) e^{-ixy} \, dx$$
$$= \text{Fourier transform of } f(x) \qquad (10.16)$$
$$= FT\{f(x)\}$$

$$f(x) = \frac{1}{2\pi} \int_{-\infty}^{+\infty} F(y) e^{+ixy} \, dy$$
$$= \text{Inverse Fourier transform of } F(y) \quad (10.17)$$
$$= IFT \; \{F(y)\}$$

10.3.2. Properties of the Fourier transform

It is useful to know a few properties about a function and its Fourier transform or the effect that various operations on a function would have on the Fourier transform of the function. We would use the symbol FT for the Fourier transform.

i. Addition

The Fourier transform of the sum of two functions is the sum of the Fourier transforms of the each function.

$$FT[f(x) + g(x)] = FT[f(x)] + FT[g(x)] \quad (10.18)$$

ii. Multiplication by a constant

$$FT[a f(x)] = a FT[f(x)] \quad (10.19)$$

Scaling

If $FT[f(x)]=F(y)$,
then

$$FT[f(ax)]=\frac{1}{|a|}F\left(\frac{y}{a}\right) \qquad (10.20)$$

for a a real constant.
Also,

$$FT\left[\frac{1}{|a|}f\left(\frac{x}{a}\right)\right]=F(ay) \qquad (10.21)$$

Shifting
If $FT[f(x)]=F(y)$,
then

$$FT[f(x\pm x_0)]=e^{\pm i x_0 y}F(y) \qquad (10.22)$$

and

$$FT[e^{\pm i y_0 x}f(x)]=F(y\mp y_0) \qquad (10.23)$$

iii. Products of Convolution
If $FT[f(x)]=F(y)$ and $FT[g(x)]=G(y)$

Then

$$FT[f(x) \otimes g(x)] = F(y).G(y) \quad (10.24)$$

Inversely,

$$FT[f(x).g(x)] = \frac{1}{2\pi} F(y) \otimes G(y) \quad (10.25)$$

where the symbol $\otimes$ is used to define the convolution of two functions:

$$f(x) \otimes g(x) = \int_{-\infty}^{+\infty} f(u)g(x-u)du. \quad (10.26)$$

iv. Area

The area under the curve representing a function is equal to its Fourier transform at the origin:

$$\int_{-\infty}^{+\infty} f(x)dx = F(0). \quad (10.27)$$

Conversely, the value of a function at the origin is equal to $\frac{1}{2\pi}$ times the area of its Fourier transform:

$$f(0) = \frac{1}{2\pi} \int_{-\infty}^{+\infty} F(y)dy. \quad (10.28)$$

v. Derivation

If $FT[f(x)] = F(y)$

then

$$FT[\frac{d^m}{dt^m} f(x)] = (iy)^m F(y) \qquad (10.29)$$

Inversely,

$$FT[(-ix)^m f(x)] = \frac{d^m}{dy^m} F(y) \qquad (10.30)$$

Examples of the Fourier transformation using *Wolfram Mathematica 7.0*

Plot[Exp[-0.04 x²],{x,-10,10}, AxesLabel□{x, Exp[-0.04 x²]}]

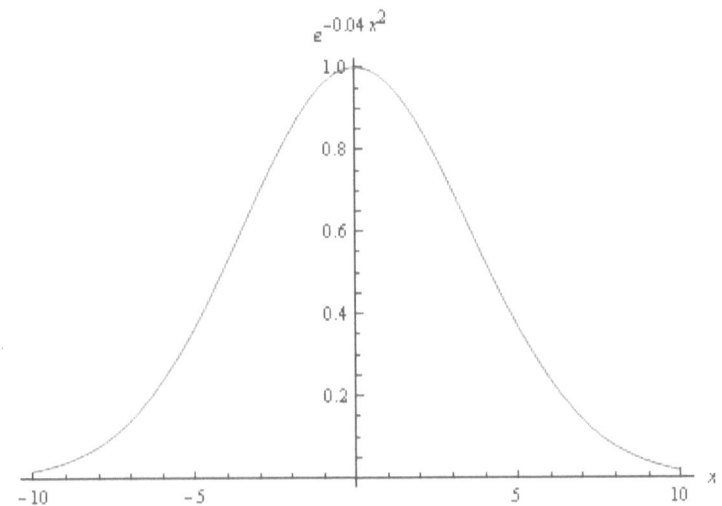

Figure 10.3. Graph of [Exp[-0.04 x^2]

FourierTransform[Exp[-0.04 x^2],x,y]

Plot[%,{y,-1,1},AxesLabel□{y,
FourierTransform[Exp[-0.04 x2]]}]

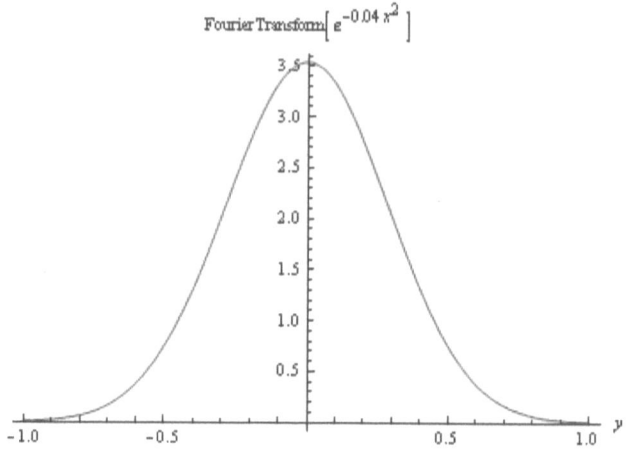

Figure 10.4. Graph of the Fourier transform of [Exp[-0.04 x²]]

10.4. Fourier transform spectroscopy

Until now the incident light is monochromatic. In other word, light with only one wavelength or one frequency. However, when the light source produces a spectrum of several frequencies, Fourier Transform then become useful.

If the frequency of wave is $v = \dfrac{c}{\lambda_i}$ then the amplitude of the superposition wave gives:

$u = \sum_{i=1}^{p} A_i \cos[2\pi v x_2]$ and the intensity at the detector is $I(y) = \sum_{i=1}^{p} \left\{ B_i \cos\left(2\pi v_i \frac{y}{2}\right) \right\}^2$ where y, is the optical path difference produced by the 2 mirrors.

By changing $\cos^2 \alpha$ into $\frac{1}{2}[\cos 2\alpha + 1]$ we obtain another appropriate form of the light intensity $I(y) = \sum_{j=1}^{p} 2C^2 j\{1 + \cos(2\pi v_j y)\}$. In this case it is understood that we consider a discrete (discontinuous) case of wavelengths λ_i $i = 1,2,3,....,n$ with n a positive integer.

Fourier Transform integral

The continuous distribution of frequency: Fourier Transform Integral.

For the case of continuous frequency distribution the intensity becomes

$$J(y) = 2\int_0^\infty G(v)(1 + \cos 2\pi v y)d\vartheta \quad \text{Where} \quad C_j^2 \text{ is}$$

being replaced by $G(v)$. The Summation sign $\sum$ is replaced by the integral sign $\int$.

In a situation where the frequency ranges from $-\infty$ to $+\infty$ let us write $J(y)$. We know that $\cos x = \Re(e^{ix})$

$$J(y) = \int_{-\infty}^{+\infty} G(v)\left[1 + e^{i2\pi v y}\right]dv. \text{ We can split } J(y)$$

$$J(y) = \int_{-\infty}^{+\infty} G(v)d(v) + \int_{-\infty}^{+\infty} G(v)e^{i(2\pi v y)}dv$$

For $y = 0$

$$J(0) = 2\int_{-\infty}^{+\infty} G(v)dv \quad \Rightarrow$$

$$J(y) = \frac{1}{2}j(0) + \int_{-\infty}^{+\infty} G(v)e^{(2\pi v y)}dv$$

$$J(y) - \frac{1}{2}J(0) = \int_{-\infty}^{+\infty} G(v)e^{i(2\pi v y)}dv \qquad (10.31)$$

This is where the Fourier integral transform comes to play a key role.

The Fourier transform for help to obtain one of the dual function $G(v)$ and $S(y)$. According to Fourier transform we can calculate $G(v)$:

$$G(v) = \int_{-\infty}^{+\infty} S(y) e^{-i(2\pi v y)} dy \qquad (10.32)$$

Problems

10.1. Compute the Fourier coefficients of the function $f(x) = x^3 + \pi x^2$

10.2. Compute the Fourier coefficients of the function $f(x)$ and its Fourier approximates in the interval $-3 < x < 3$ for N=1. Plot

$$f(x) = \begin{cases} 0 & \text{if } -3 < x < -1 \\ \cosh(x) & \text{if } -1 < x < 1 \\ 0 & \text{if } 1 < x < 3 \end{cases}$$

10.3. Prove the Euler Formula for the Fourier series represented in the above definition by

$$f(t) = a_0 + \sum_{n=1}^{\infty} [a_n \cos(n\omega t) + b_n \sin(n\omega t)], \quad \text{in other}$$

words prove the Euler formulas:

$$a_0 = \frac{1}{T} \int_0^T f(t) \, dt$$

$$a_n = \frac{2}{T} \int_0^T f(t) \cos(n\omega t) \, dt \text{ for } n \geq 1$$

$$b_n = \frac{2}{T} \int_0^T f(t) \sin(n\omega t) \, dt \text{ for } n \geq 1$$

Chapter 11
Laser optics

11.1. Introduction and history of laser

11.2. Theory and definition of laser

11.3. Spontaneous emission

11.4. Stimulated emission

11.5. Absorption

11.6. Laser idea and population inversion

11.7. Pumping schemes

11.8. Different types of lasers

11.8.1 Solid state lasers

11.8.1.1. The ruby laser

11.8.1.2. The neodymium laser

11.8.2. Gas lasers and excimers

11.8.2.1. Helium-Neon laser

11.8.2.2. Excimer laser

11.9 Light scattering by molecules: Raleigh scattering

11.10. Light scattering by particulates: Mie scattering

11.11. Inelastic scattering: Raman scattering

11.1. Introduction and history of laser

The principle of laser was first introduced in 1917 when physicist Albert Einstein described the theory of stimulated emission of radiation. In his paper "Zur Quantentheorie der Strahlung" ("On the Quantum Theory of Radiation"), Einstein established the foundation for the invention of the laser and its predecessor, the maser, in a groundbreaking rederivation of Max Planck's law of radiation based on the concepts of probability coefficients (which was later termed Einstein coefficients) for the absorption, spontaneous emission, and stimulated emission of electromagnetic radiation. In 1928, Rudolph W. Landenburg confirmed the existence of stimulated emission and negative absorption. Eleven years later, in 1939, Valentin A. Fabrikant predicted the use of stimulated emission to amplify "short" waves. In 1947, American physicist Willis E. Lamb, who won the Nobel Prize in Physics in 1955 for his discoveries concerning the fine structure of the hydrogen spectrum, and R. C. Retherford found apparent stimulated emission in hydrogen spectra and made the first demonstration of stimulated emission. In 1950, French physicist Alfred Kastler, winner of the Nobel Prize for Physics in 1966 for his discovery and development of optical methods for studying Hertzian resonances in atoms, proposed the method of optical pumping, which was experimentally confirmed by Brossel, Kastler, and Winter two years later.

Overall, during the 1920s, '30s, and '40s, physicists were preoccupied by the new discoveries of quantum mechanics, particle physics, and nuclear physics. For the most part, the possibility of laser action lay dormant, although the needs of science and technology for such a device grew. During World War II, experience gained in the development of radar and the continuation of such work at higher microwave frequencies prompted scientists to explore the conditions that were necessary for laser action to be achieved.

Working at Columbia University in the 1950s, Charles Townes and Arthur Schawlow discovered a microwave device that amplified radiation by stimulation emission of radiation. The device was termed MASER, an acronym for Microwave Amplification by the Stimulation Emission of Radiation. During the remainder of the fifties, the maser principle was employed in many materials. In

1958, Townes and Arthur L. Schawlow published an important paper in which they discussed the extension of maser principle to the optical region of the electromagnetic spectrum. They developed the concept of an optical amplifier surrounded by an optical mirror resonant cavity to allow for growth of the beam. Townes and Schawlow both received Nobel Prizes for their work in this field. In 1960, Theodore Maiman of Hughes Research Laboratory produced the first laser using a ruby crystal as the amplifier and a flash lamp as the energy source. Many ideas for laser applications quickly followed. However, the term *laser* was first introduced to the public in Gould's 1959 conference paper "The LASER, Light Amplification by Stimulated Emission of Radiation." Gould intended *-aser* to be a suffix, to be used with an appropriate prefix for the spectrum of light emitted by the device (X-rays: xaser, ultraviolet: uvaser, etc.) None of the other terms became popular, although *raser* was used for a short time to describe radio-frequency emitting devices. Gould's notes included possible applications for a laser, such as spectrometry, interferometry, radar, and nuclear fusion. He continued working on his idea and filed a patent application in April 1959. The U.S. Patent Office denied his application and awarded a patent to Bell Labs in 1960. This sparked a legal battle that ran twenty-eight years, with scientific prestige and much money at stake. Gould won his first minor patent in 1977, but it was not until 1987 that he could claim his first significant patent victory when a federal judge ordered the government to issue patents to him for the optically pumped laser and the gas discharge laser.

In 1961, Ali Javan, W. Bennett, and D. Harriott of Bell Laboratories developed the first gas laser using a mixture of He and Ne gases. At the same laboratories, L. F. Johnson and K. Nassau demonstrated the first neodymium laser, which has since become one of the most reliable lasers available. This was followed in 1962 by the first semiconductor laser, demonstrated by R. Hall at the General Electric Research Laboratories. In 1963, C. K. N. Patel of Bell Laboratories discovered the infrared carbon dioxide laser, which is one of the most efficient and powerful lasers available today. The same year, E. Bell of Spectra Physics discovered the first ion laser, in mercury vapor. In 1964, W. Bridges of Hughes Research Laboratories discovered the argon ion laser, and in 1966, W. Silfvast, G. R. Fowles, and B. D. Hopkins produced the first blue helium-cadmium metal vapor laser.

Thus, a myriad of laser types have been discovered since 1960. The first well-known rare gas—halide excimer—laser was observed in xenon fluoride (XeF) by J. J. Ewing and C. Brau of the Avco-Everett Research Laboratory in 1975. Since then, many lasers have been discovered to improve lasers' operation, control, and reliability.

11.2. Theory and definition of laser

The laser light, like the light of any ordinary light bulb, is emitted when atoms make a transition from a quantum state of a given energy to a quantum state of lower energy. However, in a laser, the atoms act together to produce light with the following special characteristics:

a- Laser light is monochromatic.

b- Laser light is coherent.

c- Laser light is sharply focused.

The word *laser* is an acronym for Light Amplification by the Stimulated Emission of Radiation. A laser exploits three fundamental phenomena that occur when an electromagnetic (e.m.) wave interacts with a material, namely, the process of spontaneous emission, stimulated emission, and the process of absorption.

11.3. Spontaneous emission

Let us consider two energy levels: 1 (lower state) and 2 (upper state) of some given material with respective energies E_1 and E_2 ($E_1 < E_2$). We assume level 1 to be the ground level. If an atom (or molecule) of the material is initially in level 2, since $E_2 > E_1$, the atom will tend to decay to level 1 at random. The corresponding difference of energy $E_2 - E_1$ will therefore be released by the atom. When this energy is delivered in the form of an e.m. wave, the process is called *spontaneous (or radiative)* emission because the event was not triggered by any outside influence (see Figure 1). The frequency of the radiated wave is then given by the Planck relation:

$$\nu = (E_2 - E_1)/h \qquad (11.1)$$

where h is Planck constant: $h = 6.6256 \times 10^{-34}$ Js

Thus, spontaneous emission is characterized by the emission of a photon of energy:

$$hv = E_2 - E_1 \qquad (11.2)$$

When the energy is delivered in the form other than an e.m. radiation, the decay will be called nonradiative decay.

If at time t, the number N_2 is the number of atoms in level 2, the rate of decay of these atoms is obviously proportional to N_2:

$$\left(\frac{dN_2}{dt}\right)_{sp} = -AN_2 \qquad (11.3)$$

A is the spontaneous emission probability and has the dimension of $(time)^{-1}$

$\tau_{sp} = \dfrac{1}{A}$ is the spontaneous emission lifetime

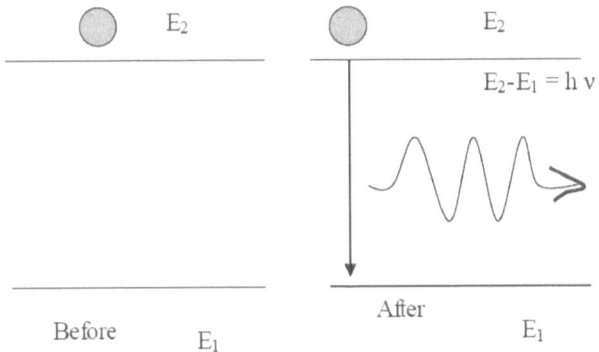

Figure 11.1. Conceptual diagram of a spontaneous emission with a two-level energy system

11.4. Stimulated emission

One can assume that the atom was initially at level 2 and that an e. m. wave of frequency $v = \dfrac{E_2 - E_1}{h}$ is incident on the material. This e. m. wave will force the atom to undergo the transition from level 2 to level 1. In this case, a difference of energy $E_2 - E_1$ is delivered in the form of an e. m. wave, which adds to the incident one. This is the phenomenon of stimulated emission (see Figure 2). There is a fundamental distinction between the spontaneous and the stimulated emission processes. In the case of spontaneous emission, the event was not triggered by a external agent. Also the wave can be emitted in all directions. The random creation of photons results in waves of random phase and the light is said to be incoherent. In the case of stimulated emission the incoming wave determines the direction of the emitted wave. The photon produced by stimulated emission is of almost equal energy to that which caused stimulated emission and hence the light waves associated with must be at nearly the same frequency. Also, the process can be characterized by means of the following equation:

$$\left(\frac{dN_2}{dt}\right)_{st} = -W_{21}N_2 \qquad (11.4)$$

where $\left(\frac{dN_2}{dt}\right)_{st}$ is the rate at which transition from level 2 to level 1 occurs as a result of stimulated emission and W_{21} is called the stimulated transition probability. The coefficient W_{21} has the dimension of $(time)^{-1}$. However, W_{21} depends on the intensity of the incident electromagnetic wave. More precisely, for a plane e. m. wave, it will be shown that:

$$W_{21} = \sigma_{21} F \qquad (11.5)$$

where F is the photon flux of the incident wave, and σ_{21} (the stimulated-emission cross section), has the dimension of area. F can be computed using the formula:

$$F = \frac{I}{\hbar \omega} \qquad (11.6)$$

where, I is the intensity of the incident e. m. wave, ω is the e. m. wave angular frequency, and

$$\hbar = \frac{h}{2\pi} \qquad (11.7)$$

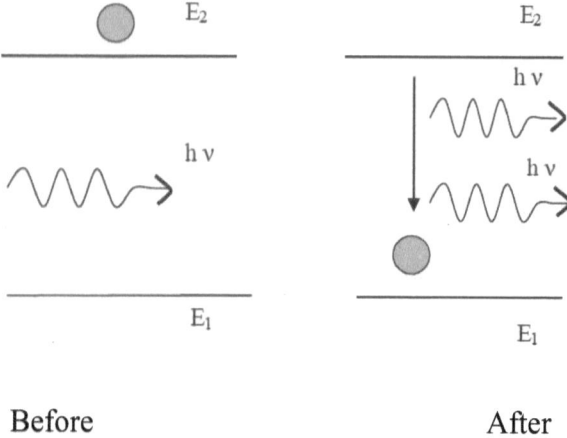

Before   After

Figure 11. 2. Conceptual diagram of the stimulated emission process

11.5. Absorption

It can be assumed that the atom is initially lying in the level 1. As this is the ground level, the atom will remain in that level unless some external stimulus is applied to it. It can also be assumed that an e.m. wave of frequency v is again applied incident on the material. In this case there is a finite probability that the atom raises to level 2. The energy difference $E_2 - E_1 = h\gamma$, required for the transition is obtained from the incident e. m. wave. This is the Absorption process (see Figure 3). One can also define here an absorption rate by means of the equation:

$$\frac{dN_1}{dt} = -W_{12}N_1 \qquad (11.8)$$

N_1 is the number of atoms (per unit volume), which, at the given time are lying in level 1.

Furthermore,

$$W_{12} = \sigma_{12}F \qquad (11.9)$$

where, W_{12} is the absorption probability, and σ_{12} is the absorption cross-section:

$\sigma_{12} = \sigma_{21} = \sigma$ as shown by Einstein showed in the beginning of the century.

σ is referred to as the transition cross-section. The number of atoms per unit volume in some given level is called population of that level. The probability of stimulated emission and absorption are equal.

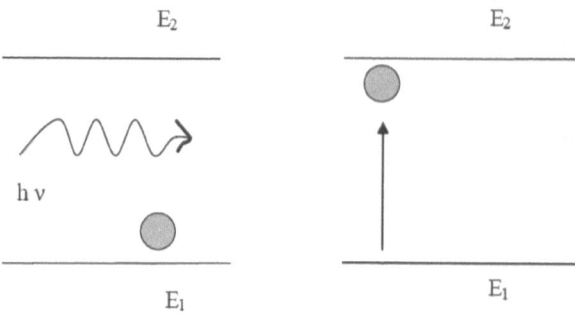

Figure 11.3. Conceptual diagram of the absorption

11.6. Laser idea and population inversion

Consider two arbitrary energy levels 1 and 2 of a given material and let N_1 and N_2 be their respective populations (Figure 4). For a plane wave with an intensity corresponding to a photon flux traveling along the z-direction in the material, the elemental change dF of the flux due to both stimulated emission and absorption is given by[50]:

$$dF = \sigma F(N_2 - N_1)dz \qquad (11.10)$$

Therefore, if $N_2 > N_1$, $dF > 0$, the material behaves as an amplifier, whereas if $N_2 < N_1$, $dF < 0$, the material behaves as an absorber.

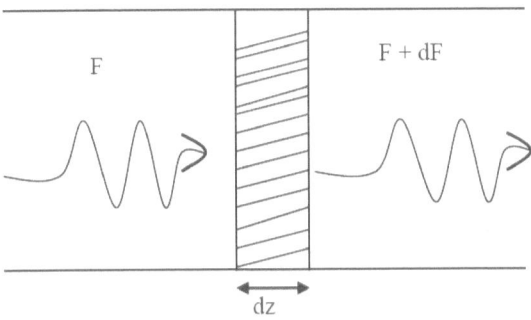

Figure 11.4. Photon flux traveling in a material of thickness dz

[1]It is known that in the case of thermal equilibrium the energy
level populations are described by Boltzmann's statistics.
Thus, if N_1^e and N_2^e are the thermal equilibrium populations
of the two levels, we have:

$$\frac{N_2^e}{N_1^e} = \exp\left[-\frac{(E_2 - E_1)}{kT}\right] \qquad (11.11)$$

where k is the Boltzmann's constant, and T the absolute temperature of the material. In consequence, for the case of thermal equilibrium, $N_2 < N_1$; therefore, dF < 0 and the material acts as an absorber. This is what happens under ordinary conditions. If however a non-equilibrium conditions is achieved for which $N_2 > N_1$, then dF > 0 and the

---

[1] *Principles of lasers second edition, Orazio Svelto and David C. HannPlenum Press New York, 1982*

material will behave as an Amplifier. In this case, there exists a population inversion in the material. A material having a population inversion is generally to referred as an active material. If the transition frequency ν falls in the microwave region, this type of amplifier is called a maser amplifier. The word maser is an acronym for Microwave Amplification by the Stimulated Emission of Radiation. Conversely, if the transition frequency ν falls in the optical region the amplifier is called Laser amplifier. Similar terminology is used not only for visible light by also for the far- or near-infrared, the ultra-violet and even the x-ray region.

To make an oscillator from an amplifier, it is necessary to introduce a suitable positive feedback. In the microwave range, this is done, by placing the active material in a resonant cavity having a resonance at the frequency ν. In the case of Laser, the feedback is often obtained by placing the active material between two highly reflecting mirrors. In this case a plane e. m. wave traveling in a direction orthogonal to the mirrors will bounce back and forth between the two mirrors and be amplified on each passage through the active material. If one of the mirrors is made partially transparent, a useful output beam can be extracted. It is useful to realize that the threshold condition must be fulfilled for the laser action to happen. Indeed the oscillation will start when the gain of the active material compensates the losses in the laser.

The gain per pass[50] in the active material is given by solving equation:

$$\frac{F}{F_0} = \exp[\sigma(N_2 - N_1)\ell] \qquad (11.12)$$

where, F is the output photon flux, F_0 the input photon flux, and ℓ the thickness of the active material. If the only losses present in the cavity are those due to the transmission losses, the threshold will be reached when[50]

$$R_1 R_2 \exp[2\sigma(N_2 - N_1)\ell] = 1 \qquad (11.13)$$

where, R_1 and R_2 are the power reflectivity of the two mirrors. Therefore, the threshold is reached when the population inversion reaches a critical value $(N_2 - N_1)_c$ known as the critical inversion[50] and given by:

$$(N_2 - N_1)_c = -\frac{\ln(R_1 R_2)}{2\sigma\ell}$$

$$(11.14)$$

At this stage, an oscillation will build up from the spontaneous emission. The photons which are spontaneously emitted along the cavity axis will, in fact, initiate the amplification process. This is the basis of a laser oscillator, or laser, as it is simply known. Figure 5 shows a schematic of a typical laser with the optical resonator.

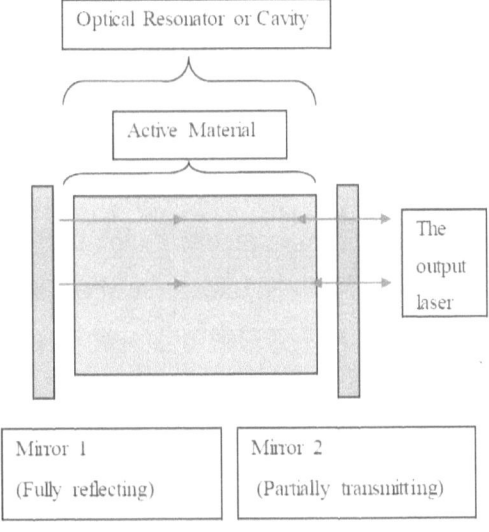

Figure 11.5 Simplified schematic of a typical laser

11.7. Pumping schemes

Considering the problem of how a population inversion can be produced in a given material, at first sight, it might seem that it would be possible to achieve this through the interaction of the material with a sufficiently strong e.m. field at the frequency v. At thermal equilibrium, as level 1 is more populated than level 2, absorption will in fact predominate over stimulated emission. The incoming wave would produce more transitions from level 1 to level 2 than from level 2 to level 1, and would hopefully in this way end-up with a population inversion. It may be seen immediately, however that such a system would not work (at least in the steady state).

When in fact the condition is reached such that the populations are equal ($N_2 = N_1$), then the absorption and stimulated process will compensate one another and, the material will then be transparent. This situation is often referred to as two-level saturation. With the use of just two levels, 1 and 2, it is impossible to produce a population inversion. It is then natural to question whether this is possible by some suitable use of more than two levels out of the infinite set of levels of the given atomic system. As it can be seen, the answer in this case is positive, and a discussion will be made of a three- or four-level laser, depending upon the number of levels used. In a three-level laser, the atoms are in some way raised from the ground level 1 to level 3. If the material is such that, after an atom has been raised to level 3, it decays rapidly to level 2, then in this way a population inversion can be obtained between levels 2 and 1. In a four-level laser, atoms are again raised from the ground level 1 to level 4. If the atom then decays rapidly to level 3, a population inversion can again be obtained between levels 3 and 2. Level 3 is not stable.

It is a meta-stable level. Once an oscillation starts in such a four-level laser, however, the atoms will then be transferred to level 2 (due to stimulated emission). For cw (continue wave) operation of a four-level laser, it is necessary that the transition from level 2 to level 1 should also be very fast.

Now, it has been seen how one can use three or four levels of a given material to produce population inversion. Whether a system will work in a three- or four-level scheme (or whether it will work at all!) depends on whether the various conditions given above are fulfilled. A question could be why one should be bothered with a four-level scheme when a three level scheme already seems to offer a suitable way of producing a population inversion. The answer is that one can in general produce a population inversion much more easily in a four-level than in a three level laser. To observe this, it can be noticed that the energy difference between the various levels are usually much greater than kT. According to Boltzmann statistics, all atoms are initially (at equilibrium) in the ground level. If N_t is the total number of atoms per unit of volume of material, these will initially all be in level 1 for the three level case. Let us now begin by raising atoms from level 1 to level 3. They will then decay to level two, and if this decay is sufficiently fast, level four will remain more or less empty. In this case, it is interesting to first raise half of the total population N_t to level 3 in order to equalize the population of levels 2 and 3. From this point on, any other atom, which, is raised will then contribute to population inversion. In a four-level laser, however, since level 2 is also initially empty, any atom, which has been raised, is immediately available for a population inversion. The above discussion shows that whenever possible, an observation should be made for a material which can operate as a four-level system rather than a three-level system. The use of more than four levels is, of course also possible. The process by which atoms are raised from level 1 to level 3 (in a three level scheme) or from 1 to 4 (in a four level scheme) is known as pumping. There are several ways in which this process can be realized in practice e.g. if light of sufficient intensity is available or if an electrical discharge takes place in an active medium.

11.8. Different types of lasers
11.8.1 Solid state lasers
11.8.1.1. The ruby laser

The advent of maser triggered the speculation that such a device might made to operate in the visible region of the spectrum. But in 1958, Schawlow and Townes suggested that the resonant cavity might take the form of two parallel plane mirrors facing each other some distance apart. This form of resonator is well known as Fabry - Perot interferometer. In 1960 Maiman of the Hughes Research Laboratory obtained the pulsed laser action at $6943 \, \overset{o}{A}$ in the red region of the spectrum using a ruby crystal as the active material. One of the ends of the ruby crystal was made almost totally reflecting and the other one about 10% transmitting or 90% reflecting in order to obtain some output from the device. The ruby laser of Maiman is a three level system. The ruby mineral (corundum) is aluminum oxide (Al_2O_3) with a small amount (about 0.05%) of chromium (Cr^{3+}) which gives it its characteristic pink or red color by absorbing green and blue light. After receiving a pumping flash from the flash tube, the laser light emerges for as long as the excited atoms persist in the ruby rod, which is typically about a millisecond. Ruby laser wavelength is $\lambda_{ul}=694.3\ nm$. The laser transition probability is $A_{ul}=333/s$ while the upper laser level lifetime is $\tau_u=3\ ms$. The stimulated emission cross section is $\sigma_{ul}=2.5\times10^{-24}\ m^2$. While the index of reflection is the gain medium is approximately 1.76 the doping density is 0.05 % by weight. Generally operated in the room temperature, the thermal conductivity of the ruby laser rod is 42 W / m-K at 300 K. Thermal expansion coefficient of the laser rod is $5.8\times10^{-6}/°C$. The pumping method is the flash lamp, the pumping bands 404 nm and 554 nm with bandwidths of 50 nm each. In normal circumstances the laser output power is up to 100 J / pulse. The mode can be a single mode or a multi – mode. As applications, the ruby laser is used in holography. It is used in plasma physics to measure plasma properties such as electron density and temperature. Ruby laser is used to remove tattoos and skin lesions resulting from excess in melanin. A pulsed ruby laser was used for the famous laser ranging experiment which was conducted with a corner reflector placed on the moon by the Apollo astronauts. This determined the distance to the Moon with an accuracy of about 15 cm. A schematic diagram of a ruby laser is given in Figure 6 and pumping Levels for the ruby are shown in Figure 7.

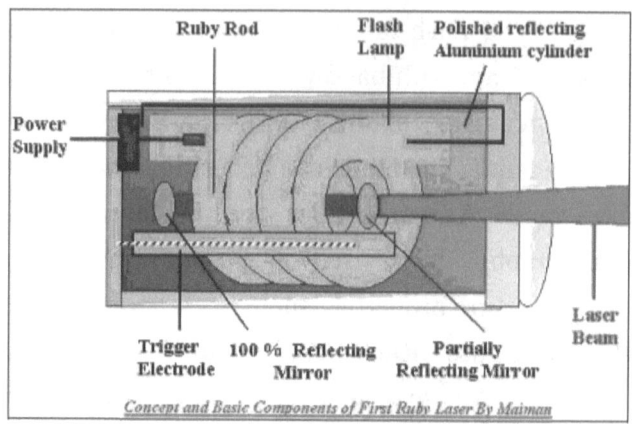

Figure 11.6. Schematic diagram of a Ruby Laser

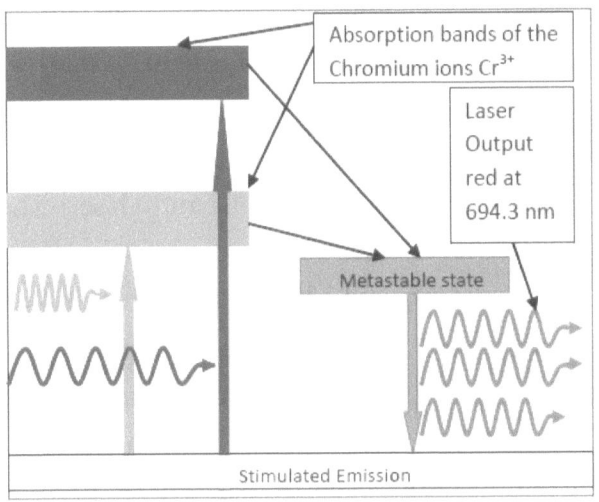

Figure 11.7. Pumping Levels for Ruby Laser

11.8.1.2. The neodymium laser

The Neodymium laser (Figure 8) is a more recently developed laser and is 4 level types (Figure 9). Two Neodymium laser have emerged: the Glass and the YAG (Yttrium Aluminum Garnet) Calcium tungsten $CaWO_4$ has also been used with success. The impurity ion is the Nd^{3+}.

When doped in YAG the Nd:YAG crystal produces laser output primarily at $1.064\ \mu m$.

When doped in Glass the Nd: Glass medium lases at wavelengths ranging from $1.054\,\mu m$ to $1.062\,\mu m$ depending upon the type of Glass used. The Nd laser incorporates a four level System and has much lower pumping threshold than the ruby laser. The upper laser level life time is relatively long, $230\,\mu s$ for Nd:YAG and $320\,\mu s$ for Nd:Glass. Nd: YAG crystal has a good optical quality a high thermal conductivity. The doping concentrations for Nd:YAG is 0.725 % by weight. We can note that the Nd: Glass laser material has a poor thermal conductivity. Now, the question one might ask is what about the laser structure? Neodymium doped lasers range from small diode-pumped versions up to high –average power with average power up to several kilowatts. There exists several laser structures. The first we would like to mention and the most useful type is the Flashlamp–pumped Q-switched oscillator amplifier system. The amplifier typically increases the oscillator output energy by up to a factor 10. The second structure of the laser is the Flashlamp – pumped cw Nd: YAG laser system. The laser is Q – switch at high repetition rate of order of 20 kHz with an average power of 15 w.

There exists also the diode pumping Nd : YAG lasers system.

Electrode   Flash tube
  Electrode

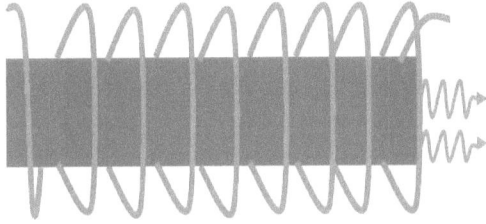

Figure 11.8. Schematic of the Nd:YAG laser. Laser Output in the Infrared at 1065 nm

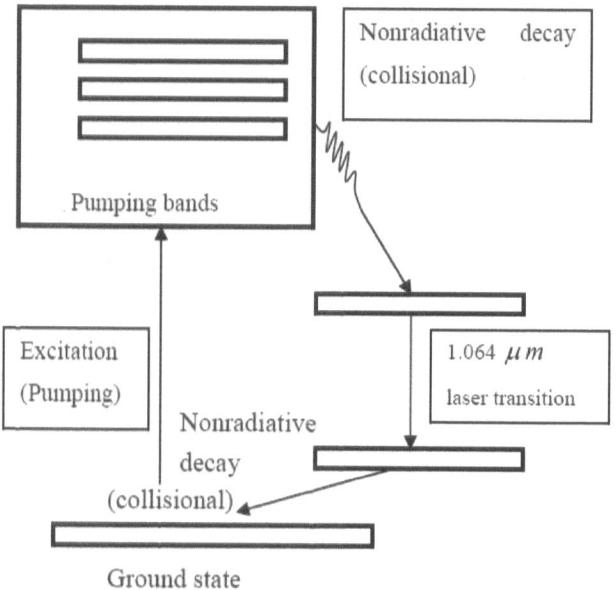

Figure 11.9. Energy levels of the Nd : YAG laser crystal

Excitation can be achieved by broadband xenon flashlamps as well as with other lasers such as the Gallium Arsenic GaAs diode laser. There is a broad range of pumping wavelength in the Nd:YAG system over the region from 0.3 μm to 0.9 μm. The output power is about 1J/pulse for the Nd: YAG laser and up to 10kJ/pulse for the Nd: Glass. This type of laser is very effective for frequency doubling the 1.064 μm light to the green at 0.532 μm and tripling to the 355 nm Nd:YAG, which, is used in our experiments. The 355 nm third harmonic generation Nd:YAG laser produces uv (ultraviolet) emission at 408 nm for water vapor, 387 nm for nitrogen, and 376 nm for oxygen.

11.8.2. Gas lasers and excimers
11.8.2.1. Helium-Neon laser
General description

Helium – Neon laser (Figures 10 & 11) is the first gas laser developed by Ali Javan, W. Bennet and D. Harriot of Bell Laboratories. It is trouble free laser and has an extremely long lifetime. Laser action occurs within the neutral atomic species. The most common wavelength is the 632.8 nm transition in the red portion of the visible spectrum. There's also some additional radiation of wavelengths 543.5 nm (green), 594 nm (yellow), 612 nm (orange) and the infrared (1.523 μm) The He- Ne laser operates with the continuous wave and produces power in the range of 0.5–50 mW in the red. It is characterized by a strong stability and low-noise output

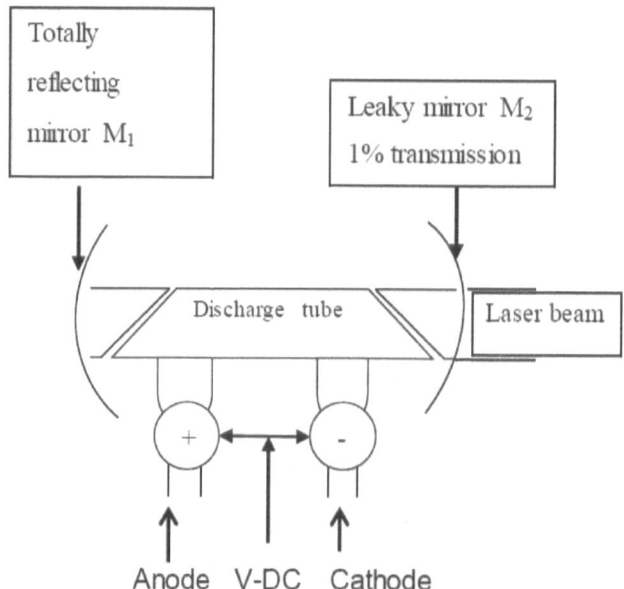

Figure 11.10. Schematic of the He-Ne laser structure

The power supply goes up to 1kV. The power emitted from a 1% transmitting mirror (approximately 99% reflecting) at the cathode end of the laser. He - Ne pressure ratio is 5:1. The operation is as follow: the glass discharge is filled with a mixture of helium and neon (15% of neon), neon being the medium where laser action occurs. An applied potential V-dc sends electrons through the discharge tube. Those electrons collide with the atoms of Helium which in turn collide with Neon atoms to trigger the light along the length of the tube. The light passes through the windows W and reflects back and forth through the tube from M_1 and M_2 to cause more neon atoms emissions. A few amount of light leaks through the mirror M_2.

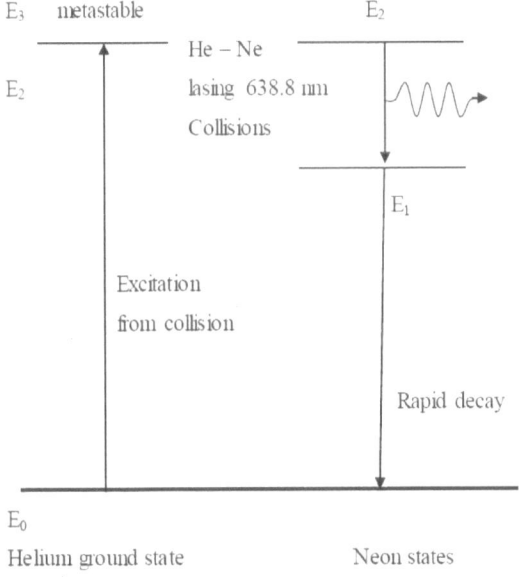

Figure 11.11. Energy level of He-Ne laser: a four level system

Interpretation and applications

The electric current form the electric charges causes the He atoms to raise to a higher level E_3 which is metastable. The energy of He at level E_3 (20.61 eV) is very close to the one of Ne at level E_2 (20.66eV). Thereafter, when the a metastable E_3 Helium atom and a ground state E_0 neon atom collide, the excitation energy of the He atom is often transferred to the Ne atom which then moves to the state E_2. As a result the Ne level E_2 becomes more populated that the Ne level E_1. Atoms in level E_1 decay rapidly to level E_0.

When atom decay from level E_2 to level 1 E_1, a coherent beam of red laser light of wavelength 632.8 nm build up rapidly and escape to form a useful external He-Ne laser beam. Finally it is useful to mention that for He-Ne laser, the pumping method is the electrical discharge, the mode of operation is cw (continuous waves) and output power can rise from 0.5 mW to 100 mW. He-Ne laser has many applications such as in interferometry, laser printing, bar code reading, pointing and directional reference beams.

11.8.2.2. Excimer laser

An important class of molecular lasers involving transitions between different electronic states is that of excimer lasers. Consider a diatomic molecule A_2 with potential energy curve as shown in Figure 12. As the ions in the ground state have the same charge, they are repulsive and the molecules formed by these ions exist only in the monomer form. Also, as the potential energy curve for the excited state has a minimum, the molecule A_2 does exist in the dimer form in the excited state. Such a molecule A_2^* is called an excimer (contraction of the words "excited dimer"). The laser produced by the transition between the upper level (bound) and the lower level (free) state is called an excimer laser. An excimer laser has three important properties due to the fact that the ground state is repulsive: (1) once the molecule, after undergoing the laser transition, reaches the ground state, it immediately dissociates, and the lower laser level will always be empty, (2) no well-defined rotational - vibrational molecules exist, and (3) the transition is broad. This allows the possibility of tunable laser radiation over the broad band transition. Tables 1 and 2 show some excimer lasers with different wavelengths for the Raleigh signal and the Raman. The values in table 2 were obtained by using the formula 73. In the case of the 248 nm KrF, excimer Laser (using Kr, F$_2$, and a buffer gas in the mixture), the following mechanism plays a very important role:

$e^- + Kr \rightarrow 2e^- + Kr^+$

$2e^- + F_2 \rightarrow F^- + F^-$

$2F^- + Kr^+ + M(Buffer\ gas) \rightarrow KrF^* + F + M$

(65)

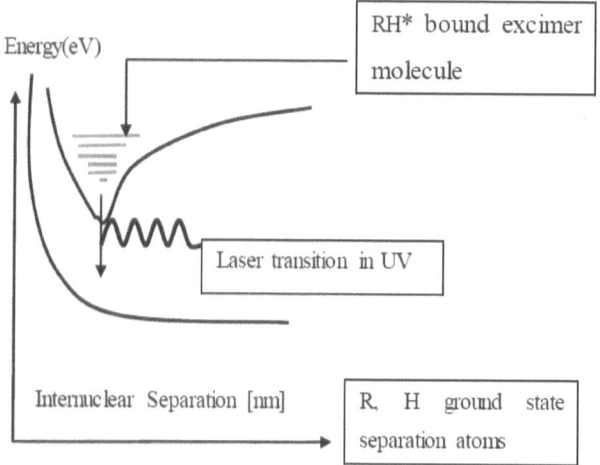

Figure 11.12. Energy levels of an excimer laser

Also, rare - gas - halide excimer lasers can be pumped by an electron beam or an electrical discharge.

Laser	Wavelengths
KrF Krypton Fluoride	248 nm
ArF Argon Fluoride	193 nm
XeF Xenon Fluoride	251 nm
XeCl Xenon Chloride	308 nm

Table 11.1. Common excimer lasers

11.9 Light scattering by molecules: Raleigh scattering

Ignoring the depolarization[48] effects and the adjustment for temperature and pressure, the molecular angular scattering coefficient $\beta_{\theta,m}$ at wavelength λ in the direction relative to the direction of the incident light can be shown to be:

$$\beta_{\theta,m} = \frac{\pi^2 (m^2 - 1)^2 N}{2 N_s^2 \lambda^4}[1 + \cos^2(\theta)] \qquad (11.15)$$

where, m is the real part of the index of refraction of the medium, and N is the number of molecules per unit of volume (number density) at the existing temperature and pressure, N_s is the number density of molecules at standard conditions ($N_s = 2.547 \times 10^{19} cm^{-3}$ at Ts = 288.15K and Ps=101.325kPa).

Integrating over all possible angles, one obtains the molecular volume scattering coefficient[48] as:

$$\beta_m = \int_{\varphi=0}^{2\pi}\int_{\theta=0}^{\pi} \beta_{\theta,m} \sin(\theta)\, d\theta\, d\varphi$$
$$= \frac{8\pi^3 (m^2 - 1)^2 N}{3 N_s^2 \lambda^4} \quad (11.16)$$

The molecular phase function[48] $P_{\theta,m}$ normalized to one is given by:

$$P_{\theta,m} = \frac{\beta_{\theta,m}}{\beta_m} = \frac{3}{16\pi}[1 + \cos^2(\theta)] \quad (11.17)$$

For the molecular scattering, the cross section defines the amount of scattering due to a single molecule. The molecular cross section σ_m is the ratio

$$\sigma_m = \frac{\beta_m}{N} \quad (11.18)$$

Therefore it is given by:

$$\sigma_m = \frac{8\pi^3(m^2-1)^2}{3N_s^2 \lambda^4} \qquad (11.19)$$

From this, the total angular and molecular scattering intensity[48] is proportional to λ^{-4}. Therefore, atmospheric gases scatter much more light in the UV region than in the IR portion of the spectrum. This kind of scattering called Rayleigh scattering is inherent not only to molecules but also to particulates, for which the radius is small relative to the wavelength of the incident light.

11.10. Light scattering by particulates: Mie scattering

As the characteristics sizes of the particulates approach the size of the wavelength of the incident light, the nature of the scattering changes dramatically. This is called Mie scattering after Mie who first provided a theoretical explanation (1908). Mie scattering holds also for situations where the size of the particles is much greater than the wavelength.

11.11. Inelastic scattering: Raman scattering

Although the dominant mode of molecular scattering in the atmosphere is elastic scattering, commonly called Rayleigh scattering, it is also possible for the incident photon to interact inelastically with the molecules.

Let a sample be irradiated by intense laser beams of frequency γ_0 in the UV- visible region (see Figure 11.13). The scattered light is usually observed in the direction perpendicular to the incident beam. The scattered light consists of two types: One, called Rayleigh scattering, is strong and has the same frequency as the incident beam (γ_0) and the other, called Raman scattering is very weak (~10^{-5} of the incident beam) and has frequencies $\gamma_0 \pm \gamma_m$, where γ_m is the vibrational frequency of a molecule.

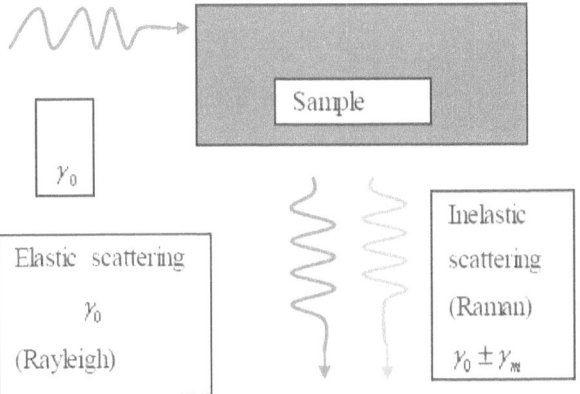

Figure 11.13. Conceptual diagram of the elastic and inelastic scattering

The $\gamma_0 - \gamma_m$, and $\gamma_0 + \gamma_m$ lines are called stokes and anti-stokes lines respectively. The incident frequency is γ_0 (Rayleigh scattering), while Raman scattering frequencies are $\gamma_0 + \gamma_m$ (anti-stokes), and $\gamma_0 - \gamma_m$ (stokes). For the case of stokes' lines the corresponding Raman wavelength is obtained by taking the shift between the incident wavelength and the wavelength corresponding to the vibrational frequency. The change in the energies between the incident and the scattered light is given by:

$$\Delta E = E_i - E_s = h\gamma_0 - h(\gamma_0 - \gamma_m) = h\gamma_m$$

Therefore, $\quad h\gamma_0 - h\gamma_m = h(\gamma_0 - \gamma_m)$

Thus, $\dfrac{1}{\lambda_0} - \dfrac{1}{\lambda_m} = \dfrac{1}{\lambda_{Ra}}$, where λ_0, λ_m, and λ_{Ra}, correspond to the incident wavelength, the vibrational wavelength, and the Raman stokes wavelength, respectively.

Molecules	Molecular molar mass (g/mole)	Normal mode of vibration Raman-active (cm^{-1})	Raman corresponding wavelengths (nm) Incident Krypton Fluoride KrF At 248 nm	Raman corresponding wavelengths (nm) Incident Krypton Fluoride KrF at 248 nm	Raman corresponding wavelengths (nm) Incident Neodymium Yag laser Nd YAG at 355 nm	Raman corresponding wavelengths (nm) Incident Neodymium Yag laser Nd YAG at 355 nm	Raman corresponding wavelengths (nm) Incident Xenon Fluoride XeF at 351 nm	Raman corresponding wavelengths (nm) Incident Xenon Fluoride XeF at 351 nm
			Stokes line	Anti-stokes line	Stokes line	Anti-stokes line	Stokes line	Anti-stokes line
Oxygen O_2	32	1555	257.96	238.79	375.79	336.42	371.26	332.83
Nitrogen N_2	28	2331	263.2	234.44	387.14	327.86	382.27	324.45
Water vapor H_2O	18	3657	272.7	222.38	407.96	314.21	402.69	311.07

Table 11.2. Comparison of Raman wavelengths with KrF (248 nm),

Nd : YAG (355 nm), and XeF (351 nm)

Chapter 12

Lidar theory

12.1. Introduction and definition of lidar

12.2. The lidar equation

12.3. Some basic definitions relative to water vapor

12.3.1. Water vapor mixing ratio

12.3.2. Relative humidity

12.3.3. Specific humidity

12.3.4. Water vapor density

12.3.5. Absolute humidity

12.3.6. Partial pressure and vapor pressure

12.3.7. The wet-bulb temperature

12.4. Overview of lidar techniques used by scientists

12.1. Introduction and definition of lidar

LIDAR is an acronym for Light Detection and Ranging. It makes use of a laser to excite backscattering in the atmosphere. This backscatter signal is observed using a telescope receiver. The optical signals received by the telescope possibly selected by wavelength are directed to photosensitive detectors, which convert the optical signal into an electrical signal. These electrical signals are recorded as a function of time by analog-to-digital converters and/or by photon counting devices and the results are stored onto a computer.

12.2. The lidar equation

The power P(R) detected in the return signal as a function of range R may be expressed as a function of 4 terms:

$$P(R) = K\, G(R)\, \beta(R)\, T(R) \qquad (12.1)$$

where, K is the system factor that summarizes the performance of the LIDAR system $G(R)$ is the range-dependent geometric factor, $\beta(R)$ is the backscatter coefficient at distance R, and $T(R)$ is the transmission term.

The system factor:

$$K = P_o \frac{c\tau}{2} A \eta \qquad (12.2)$$

where, P_0 is the average power of a single laser pulse (W), τ is the temporal pulse length in seconds, A is the area of the telescope (m^2), η is the overall LIDAR system efficiency at the laser wavelength (unitless), and c is the speed of light in vacuum.

The range dependent geometric factor:

$$G(R) = \frac{O(R)}{R^2} \qquad (12.3)$$

where O(R) is the laser beam receiver-field-of-view overlap function. The backscatter coefficient at distance R, β(R, λ) may be expressed as:

$$\beta(R,\lambda) = \sum_{j=1}^{n} N_j \frac{d\sigma_{j,sca}}{d\Omega}(\pi,\lambda) \qquad (12.4)$$

where N_j is in m^{-3}. Here, $\beta(R)$ is in $m^{-1}sr^{-1}$, and $\frac{d\sigma_{j,sca}}{d\Omega}$ is the differential backscatter cross-section in $m^2 sr^{-1}$

The transmission term:

$$T(R,\lambda) = \exp\left[-2\int_0^R \alpha(r,\lambda)\,dr\right] \qquad (12.5)$$

where $\alpha(r,\lambda)$ is the optical extinction coefficient defined as the sum of all optical transmission losses due to both scattering and absorption.

The extinction coefficient is given by:

$$\alpha(R,\lambda) = \sum_{j=1}^{n} N_j(R)\sigma_{j,ext}(\lambda) \qquad (12.6)$$

The unit of $\alpha(R,\lambda)$ is m^{-1}; $\sigma_{j,ext}$ is the total cross section   The backscatter coefficient may be expressed as

$$\beta(R,\lambda) = \beta_{molecules}(R,\lambda) + \beta_{aerosols}(R,\lambda) \qquad (12.7)$$

$\beta(R,\lambda)$ has the unit of $(m^{-1}.sr^{-1})$

The extinction coefficient α (R, λ) can be expanded in terms of the contribution of the both molecules and particles (aerosols) to the scattering and absorption of light. Thus,

$$\alpha(R,\lambda) = \alpha_{molecules,scattering}(R,\lambda) + \alpha_{molecules,absorption}(R,\lambda)$$
$$+ \alpha_{aerosols\,scattering}(R,\lambda) + \alpha_{aerosols\,absoption}(R,\lambda) \quad (12.8)$$

Lidar simple equation as expressed in the literature is given by:

$$P(R,\lambda) = P_o \frac{c\tau}{2} A\eta \frac{O(R)}{R^2} \beta(R,\lambda) \exp\left[-2\int_0^R \alpha(r,\lambda)dr\right] \quad (12.9)$$

12.3. Some basic definitions relative to water vapor
12.3.1. Water vapor mixing ratio

The water vapor mixing ratio, ω, is one of the most important atmospheric state variables and is defined as the ratio of the mass of water vapor to the mass of dry air in a given volume. It is frequently expressed in g.Kg^{-1}. If one takes 78% of dry air as the nitrogen concentration, then, the water vapor mixing ratio can be expressed as the ratio of the nitrogen Raman LIDAR signals. ω can be expressed as the ratio of the number density of water vapor N_H with respect to the number density of nitrogen N_N.

$$\omega = \frac{m_{H_2O}}{m_{dry\ air}} = \frac{MW_{H_2O}}{MW_{dry\ air}} \frac{N_{H_2O}}{N_{dry\ air}} \qquad (12.10)$$

$$\omega \approx \frac{MW_{H_2O}}{MW_{dry\ air}} \frac{N_{H_2O}}{N_N/0.78} \qquad (12.11)$$

$$\omega \approx 0.485 \frac{N_{H_2O}}{N_N} \qquad (12.12)$$

where MW_{H2O} represents the molecular weight of water vapor (18 g/mol) and $MW_{dry\ air}$ is the molecular weight of dry air (28.94g/mol) [20].

Now, consider the power P received by the telescope for water vapor and nitrogen:

$$\begin{aligned} P(R,\lambda_H) &= P_o(\lambda_L)\frac{c\tau}{2} A\eta(\lambda_H)\frac{O_H(R)}{R^2} \\ &\times \beta_H(R,\lambda_L)\exp\left[-\int_0^R [\alpha(r,\lambda_L)+\alpha(r,\lambda_H)dr\right] \end{aligned} \qquad (12.13)$$

$$P(R,\lambda_N) = P_o(\lambda_L)\frac{c\tau}{2} A\eta(\lambda_N)\frac{O_N(R)}{R^2}$$
$$\beta_N(R,\lambda_L) \times \exp\left[-\int_0^R [\alpha(r,\lambda_L)+\alpha(r,\lambda_N)]dr\right] \quad (12.14)$$

where, $P(R, \lambda_X)$ is the background-subtracted power received at the Raman-shifted wavelength for either water vapor (H) of nitrogen (N) as a function of the range R, $P_o(\lambda_L)$ is the output power of the laser at the laser wavelength λ_L, N_H is the number density of water vapor molecules, N_N is the one of nitrogen molecules, $d\sigma_X(\pi)/d\Omega$ is the Raman differential backscatter cross section at the laser wavelength λ_L, and $\eta(\lambda_X)$ is the total LIDAR receiver optical efficiency for either the water vapor or nitrogen wavelength and includes factors such as the reflexivity of the telescope and the quantum efficiency of any detector. $\beta_X(R,\lambda_L) = N_X d\sigma_X(\pi)/d\Omega$ is the Raman backscatter coefficient. The exponential factor gives the two-way atmospheric transmission, where $\alpha(R, \lambda_X)$ represents the total extinction coefficient that is due to scattering and absorption by molecules and aerosols at the specific wavelength as a function of range along the path of the laser beam.

If we take the ratio of water vapor and nitrogen signals, we get:

$$\frac{P(R,\lambda_H)}{P(R,\lambda_N)} = \frac{\eta(\lambda_H)O_H(R)N_H(R)\frac{d\sigma_H(\pi)}{d\Omega}}{\eta(\lambda_N)O_N(R)N_N(R)\frac{d\sigma_N(\pi)}{d\Omega}} \quad (12.15)$$

$$\times \exp\left[-\int_0^R [\alpha(r,\lambda_H) - \alpha(r,\lambda_N)]dr\right]$$

The exponential term expresses the difference in one-way atmospheric transmission between Raman wavelengths and is abbreviated as $\Delta T(\lambda_H, \lambda_N, R)$ with

$$\Delta T(\lambda_H, \lambda_N, R) = \exp\left[-\int_0^R [\alpha(r,\lambda_H) - \alpha(r,\lambda_N)]dr\right] \quad (12.16)$$

Also,

$$\beta_H(R,\lambda_L) = N_H \frac{d\sigma_H(\pi)}{d\Omega}, \quad (12.17)$$

$$\beta_N(R,\lambda_L) = N_N \frac{d\sigma_N(\pi)}{d\Omega}, \quad (12.18)$$

Thus,

$$\frac{P(R,\lambda_H)}{P(R,\lambda_N)} = \frac{\eta(\lambda_H)O_H(R)N_H(R)\frac{d\sigma_H(\pi)}{d\Omega}}{\eta(\lambda_N)O_N(R)N_N(R)\frac{d\sigma_N(\pi)}{d\Omega}} \Delta T(\lambda_H, \lambda_N, R) \quad (12.19)$$

and,

$$\frac{N_H(R)}{N_N(R)} = \frac{P(R,\lambda_H)}{P(R,\lambda_N)} \frac{\eta(\lambda_N)O_N(R)\frac{d\sigma_N(\pi)}{d\Omega}}{\eta(\lambda_H)O_H(R)\frac{d\sigma_H(\pi)}{d\Omega}} \Delta T(\lambda_N, \lambda_H, R) \quad (12.20)$$

So, the water vapor mixing ratio can be written as:

$$\omega = 0.485 \frac{P(R,\lambda_H)}{P(R,\lambda_N)} \frac{\eta(\lambda_N)O_N(R)\frac{d\sigma_N(\pi)}{d\Omega}}{\eta(\lambda_H)O_H(R)\frac{d\sigma_H(\pi)}{d\Omega}} \Delta T(\lambda_N, \lambda_H, R) \quad (12.21)$$

or,

$$\omega = k^*(R)\frac{P(R,\lambda_H)}{P(R,\lambda_N)} \Delta T(\lambda_N, \lambda_H, R)$$
$$(12.22)$$

where,

$$k^* = 0.485 \frac{\eta(\lambda_N)O_N(R)\frac{d\sigma_N(\pi)}{d\Omega}}{\eta(\lambda_H)O_H(R)\frac{d\sigma_H(\pi)}{d\Omega}} \quad (12.23)$$

$\Delta T(\lambda_H, \lambda_N, R)$ represents the coefficient of proportionality in the above equations.

Equation (45) proves that the water vapor mixing ratio ω, can be expressed in terms of the ratio of the signals, $\dfrac{P(R, \lambda_H)}{P(R, \lambda_N)}$.

12.3.2. Relative humidity

Relative humidity is the ratio of actual vapor pressure, at a given temperature, to the saturation value at that temperature. Thus,

$$RH = \frac{e_V}{e_S} \times 100 \qquad (12.24)$$

12.3.3. Specific humidity

The specific humidity is the ratio of the mass of water vapor in a sample to the total mass of the moist air, including both the dry air and the water vapor. As ratios of masses, both specific humidity and mixing ratio are dimensionless numbers. However, because atmospheric concentrations of water vapor tend to be at most only a few percent of the amount of air (and usually much lower), they are both often expressed in units of grams of water vapor per kilogram of (moist or dry) air.

12.3.4. Water vapor density

Water vapor density ρ_v can be calculated knowing the saturated vapor density ρ_s and the relative humidity (RH). Thus

$$\rho_V = \frac{RH \times \rho_S}{100} = \rho_{dry\ air} \times \omega \qquad (12.25)$$

$\rho_{dry\ air}$ is the density of dry air about $1.29\ Kg/m^3$

12.3.5. Absolute humidity

Absolute humidity is the same as the water vapor density, defined as the mass of water vapor divided by the volume of associated moist air and generally expressed in grams per cubic meter. The term is not much in use now.

12.3.6. Partial pressure and vapor pressure

The partial pressure of a given sample of moist air that is attributable to the water vapor is called the vapor pressure. The vapor pressure necessary to saturate the air is the saturation vapor pressure. Its value depends only on the temperature of the air. (The Clausius-Clapeyron equation gives the saturation vapor pressure over a flat surface of pure water as a function of temperature.) Saturation vapor pressure increases rapidly with temperature: the value at 90°F (32°C) is about double the value at 70°F (21°C). The saturation vapor pressure over a curved surface, such as a cloud droplet, is greater than that over a flat surface, and the saturation vapor pressure over pure water is greater than that over water with a dissolved solute.

12.3.7. The wet-bulb temperature

The wet-bulb temperature is the temperature an air parcel would have if it were cooled to saturation at constant pressure by evaporating water into the parcel. The term comes from the operation of a psychrometer, a widely used instrument for measuring humidity, in which a pair of thermometers, one of which has a wetted piece of cotton on the bulb, is ventilated. The difference between the temperatures of the two thermometers is a measure of the humidity. The wet-bulb temperature is the lowest air temperature that can be achieved by evaporation. At saturation, the wet-bulb, dew point, and air temperatures are all equal; otherwise the dew point temperature is less than the wet-bulb temperature, which is less than the air temperature.

12.4. Overview of lidar techniques used by scientists

During the last thirty years, researchers at several laboratories have demonstrated that lidar has special capabilities for the remote sensing of many different properties of the atmosphere. One of these techniques that show a great deal of promise for several applications is Raman scattering. In this book, the application of the Raman scattering techniques to obtain water vapor, nitrogen, and oxygen using the 248 nm KrF laser and the 355 Nd:YAG laser is described. The first Raman measurements of atmospheric properties with lidar were carried out in the late 1960s. Fiocco and Smullin[57] were the first to record atmospheric measurements while using lidar techniques with ruby laser. J.A. Cooney[44,53] also made measurements on the Raman component of laser atmospheric backscatter. Later, Melfi et al., in 1969, and Cooney in 1970 showed that it was possible to measure atmospheric water vapor profiles using the Raman lidar technique. In 1972, Inaba and Kobayasi on one hand, and Strauch, Derr, and Cupp on the other hand, made a significant contribution in suggesting several species that could potentially be measured using the vibrational Raman techniques. Although the early tests showed that it was possible to measure water vapor with limited accuracy, recent investigations with new techniques have shown significant improvement. Particularly, the investigations of Vaughan et al. and Whiteman et al. have demonstrated rather convincingly that the Raman technique has a high capability of making accurate water vapor measurements. In 1990, Ansmann et al[64] determined atmospheric aerosol extinction profiles from measured Raman lidar signals. In 1991, Grant applied Raman and DIAL lidar techniques to measure water vapor.

The measurements of water vapor during the daytime have been demonstrated by Renault and Capitini using the solar blind region of the UV spectrum. Their work showed that the optimum wavelength was near the fourth harmonic for the Nd:YAG laser. But at this wavelength, the measurement of N_2 and H_2O are contaminated by the absorption of ozone and SO_2 in the lower troposphere[34]. Taking into account these corrections and others, Whiteman[2], in a series of two papers in 2003, points out that errors in the order of 5% and more can be introduced into the water vapor mixing ratio calculation at high altitude and errors larger than 10% are possible for the calculations of the aerosol scattering ratio and also the aerosol backscatter coefficient and extinction-to-backscatter ratio.

Despite its abundance in the atmosphere and its importance for the climate system, many questions regarding H_2O are presently unresolved. For example, the important question of feedback between the water vapor mixing ratio in the lower and upper troposphere and surface temperature is still unanswered. Although increased temperature leads to increased moisture and further warming because of the "greenhouse" effect, it is unclear whether the warming produces more water due to further evaporation (positive feedback) or if the increased upwelling causes a drying (negative feedback).

Results of calculations by Manabe and Wetherald[65] (1967) from a radiative-convective model with constant humidity suggested that the exponential increase of absolute humidity due to the sea surface temperature rise would exert a strong positive feedback. Other analyses from complex general circulation models are generally consistent with Manabe's and Weatherald's conclusions[65] showing similarly large positive feedback. Ellsaesser (1984, 1990), on the other hand, argued that an increase in the strength of convection in the tropics would cause an increase in the Hadley Cell circulation[66]. An increase in the strength of the circulation will lead to drying or negative feedback, rather than a moistening of the upper troposphere (Ellsaesser 1984, Lindzen 1990, and Sun and Lindzen 1993). In short, as far as the greenhouse effect and climate change are concerned, it is still not clear whether thermodynamics or dynamics control tropospheric water vapor. The radiative effects of clouds are yet another complicating factor in making water vapor measurements. Consequently, our understanding of the distribution of tropospheric and lower stratospheric water vapor is not as thorough as it should be. Also, although the basic workings of the hydrological cycle are well known, there is a general lack of knowledge about the interactions between some components of the hydrological cycle such as N_2, O_2, and ozone. This is in part due to the tremendous heterogeneity of the boundary conditions and complexity of these interactions and in part due to the lack of adequate observations.

Chapter 13

Equipment and experimental procedure

13.1. Introduction

13.1.1. The transmitter system

13.1.1.1. The example of the 248 nm KrF laser system

13.1.1.2. Characteristics of the 355 nm Nd:YAG lidar system

13.1.2. The receiver

13.1.3. The detector system

13.1.3.1. The photomultiplier tube, PMT

13.1.3.2. The spectrometer

13.1.3.3. The beam acquisition electronics

13.1.3.4. The filter

13.2. Experimental procedure

13.2.1. The block diagram of the lidar system

13.2.2. The SOLEX method

13.1. Introduction

A basic LIDAR instrument is made up of a laser source, telescope, and detection system. A laser pulse is emitted into the atmosphere, scatters from the air molecules and aerosols (particulates in the air such as dust) and returns through the telescope system. The detection system collects the light, and analyzes the data.

13.1.1. The transmitter system
13.1.1.1. The example of the 248 nm KrF laser system

The transmitter system is the laser itself. Figure 13.1 is a photograph of the laser system on an optics bench. The 248 nm KrF has a power less than or equal to 30 W, with a pulse more that 10 ns. The average pulse-to- pulse stability is less than or ± 12%.

Figure 13.1. Photograph of the 248 nm KrF laser device COMPEX set up on an optics bench in the Coude building

	Compex Model	KrF	Units
Wavelength		248	nm
Pulse Energy Measured, at low repetition rate	110 Multigas	300	mJ
	110 F-Version	350	mJ
Maximum Repetition Rate	110	100	Hz
Average Power measured at maximum repetition rate	110 Multigas	25	W
	110 F-Version	30	W
Nominal Pulse Duration	Compex 100 Series	30	ns, FWHM
Beam Dimensions	Compex 100 Series	24x5-10	mm
Beam Divergence	Compex 100 Series	3x1	mrad

Table 13.1. The 248 nm KrF laser specifications:

The previous table 13.1 summarizes the specifications of the 248nm KrF laser.

13.1.1.2. Characteristics of the 355 nm Nd:YAG lidar system

The following table summarizes the characteristics of a Nd:YAG LIDAR system.

Geometry	Zenith pointing, coaxial geometry with periscope, and inclement weather window
Laser	3rd harmonic of Nd:YAG, 354.7 nm, 30 Hz, 0.5 mrad divergence, 9.5 mm beam diameter, Continuum Model 9500, 350 mJ / pulse
Beam Expander	Eye Safe for exposure to single pulse via 15X Beam Expander, with diffraction limited 25.4 mm diameter, 500 mm focal length entrance lens and two 150 mm diameter exit lenses, each with 1000 mm focal length. This expander reduces the divergence of the outgoing laser beam by a factor of 15
Transmitter Mirror	15 x 20 cm2, better than 1 µrad θ & Φ precision, measurement range ~0.1 km – ~15 km; Newport ESP 300 Motion Controller
Optical Coupling	Fiber Optic - coupled telescope receiver design, Fiber: silica, multi-mode, 300 µm dia. core, 0.22 NA, positioned at telescope focal plane

Table 13.2. The 355 nm Nd:YAG laser specifications

13.1.2. The receiver

Many receiver systems are used. One of them is the telescope as seen below

Figure 13.2. The telescope is mounted on a solid concrete base, and it is completely isolated from vibrations of the dome

13.1.3. The detector system

The detector System consists of the photomultiplier tube (PMT), the spectrometer, and the beam acquisition electronics. Filters are also part of the detector system.

13.1.3.1. The photomultiplier tube, PMT

Photomultiplier tubes convert photons to an electrical signal. They have a high internal gain and are sensitive detectors for low-intensity applications such as fluorescence spectroscopy. A PMT consists of a photocathode and a series of dynodes in an evacuated glass enclosure. When a photon of sufficient energy strikes the photocathode, it ejects a photoelectron due to the photoelectric effect. The photocathode material is usually a mixture of alkali metals, which make the PMT sensitive to photons throughout the visible region of the electromagnetic spectrum. The photocathode is at a high negative voltage, typically -500V to -1,500V. The photoelectron is accelerated toward a series of additional electrodes called dynodes. These electrodes are each maintained at successively less negative potentials. Additional electrons are generated at each dynode. This cascading effect creates 10^5 to 10^7 electrons for each photoelectron that is ejected from the photocathode. The amplification depends on the number of dynodes and the accelerating voltage. This amplified electrical signal is collected at an anode at ground potential, which can be measured. Figure 13.3 presents a conceptual diagram of a PMT. The spectral response of a photomultiplier tube is usually expressed in terms of quantum efficiency (QE) or radiant sensitivity as a function of wavelength. The quantum efficiency of a PMT is the number of electrons ejected by the photocathode divided by the number of incident photons. It is customary to present the QE in percentage. Moreover, the radiant sensitivity (S) is the photoelectric current from the photocathode, divided by the incident radiant power at a given wavelength, expressed in amperes per watt (A/W).

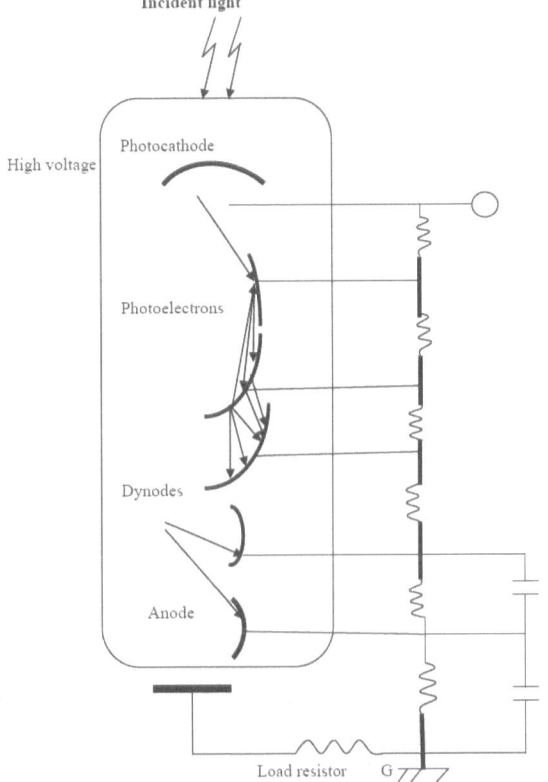

Figure 13.3. A schematic diagram of a photomultiplier with eight dynodes; electrons released from the photocathode are accelerated toward the next dynode.

The PMT used for the filter is the head-on type (transmission mode), 13 mm (1/2'') diameter. The photomultiplier tube, Hamamastu R 759 is used for the filter. It possesses a 10-stage linear focus dynode chain, which exhibits high anode sensitivity, low dark current and high stability.

The photocathode material is Cs-Te, which is sensitive to vacuum UV and UV rays but not to visible light and therefore called solar blind. Cs-Te is quite insensitive to wavelengths longer than 320 nm and the maximum allowed anode to cathode voltage is − 1250 V DC. Therefore the maximum average anode current is 0.01 mA. In the time response, the rise time is 2.5 ns and the electron transit time is 24 ns. The window material is in synthetic quartz, which transmits ultraviolet radiation down to 160 nm and offers lower absorption in the UV range. The spectral response of the detector ranges from 160 nm to 320 nm with a maximum quantum efficiency, QE, of 20% at a peak wavelength of 240 nm. Although the transmission curve is constant in the region of interest (240 nm-270 nm), the intensity of the transmitted photons rapidly changes with the applied high voltage vs the intensity of the PMT. The PMT used for the spectrometer is also a head-on type (transmission mode), 13 mm (1/2") diameter.

The photomultiplier tube, Hamamastu R1463, can be used for the spectrometer. It also possesses a 10-stage linear focus dynode chain, which exhibits High anode sensitivity, low dark current and high stability. The photocathode material is a Multialkali Na-K-Sb-Cs, which offers a high, wide spectral response from the ultra violet to near infrared region. The maximum allowed anode to cathode voltage is –1250 V DC with a corresponding maximum anode current of 0.03 mA. In the time response, the rise time is 2.5 ns and the electron transit time is 24 ns. The window material is a UV transparent, which transmits ultraviolet radiation down to 185 nm and offers lower absorption in the UV range. The spectral response of the detector ranges from 185nm to 850nm with a maximum quantum efficiency, QE, of 25% at a peak wavelength of 420 nm. The transmission curve is constant in the region of interest (240 nm-270 nm). However, the intensity of the transmitted photons rapidly changes with the applied high voltage vs the intensity of the PMT.[55] Table 13.3 shows the characteristics of a head-on type PMT. Figure 13.4 shows a plot of the spectral response of several types of photomultipliers. Numbers indicate the types of photocathode materials: 100 M, 200 M, 200 S, 500 U, 500 S, and so forth.

13.1.3.2. The spectrometer

The example of spectrometer that I will give here is the prism spectrometer (Beckman), with a 2-meter, double-folded optical path and a variable slit width is placed at the image plane of the telescope. The spectrometer is mounted at the rear of the telescope, with the slit at the focal point. The calibrated scale, which enables the observer to read the wavelengths, is visible at the top of the spectrometer. A PMT (Hamamatsu, R 1463) is mounted in a fixed position on the spectrometer. The spectrometer is aligned on the optical axes of the telescope, with the slit in the focal plane, where observations the spectral lines of mercury are made. Figure 13.5 shows a picture of the Beckman spectrometer on a table.

	Type No.	Type No.
	R 759	R 1463
Remarks	For UV range	Multialkali photocathode for UV to near IR range
Spectral Response Curve Code	200S	500U
Spectral Response Range (nm)	160 to 320	185 to 850
Spectral Response Peak Wavelength (nm)	240	420
Photo-Cathode Material	Cs-Te	MA
Window Material	Q	U
Dynode structure/No of stages	L/10	L/10
Maximum Ratings: Anode to cathode Voltage (Vdc)	1250	1250
Maximum Ratings: Average Anode Current (mA)	0.01	0.03
Cathode Sensitivity Luminous Min. (µA/lm)	-	80
Cathode Sensitivity Luminous Typ. (µA/lm)	-	120
Cathode Radiant Sensitivity (mA/W)	28b	51
Anode to Cathode Supply Voltage (Vdc)	1000	1000
Anode Radiant Sensitivity Typ. (A/W)	1.4×10^4	5.1×10^4
Gain Typ.	5.0×10^5	1.0×10^6
Rise Time Response (ns)	2.5	2.5
Electron Transit Time Response (ns)	24	24
Anode Dark Current (After 30 mn) Typ. (nA)	0.3	4
Anode Dark Current (After 30 mn) Max. (nA)	1	20

Table 13.3. Head-on type photomultiplier tubes

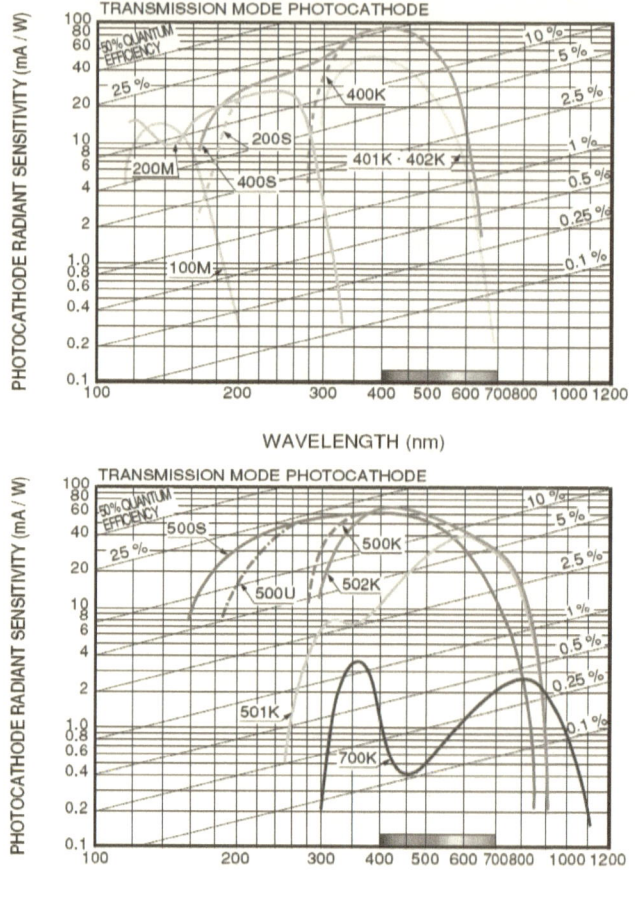

Figure 13.4. Schematic of Semitransparent photocathode spectral response characteristics.

13.1.3.3. The beam acquisition electronics
Two fast preamplifiers are used to boost very fast signals from the photomultiplier. They are connected to the output of the PMT. The preamplifiers that are used in our experiments are high-performance, non-invertible single-channel units (EG&G ORTEC, Model C-VT 120), with a wide bandwidth of 10–350 MHz, a gain of 20, rise time less than or equal to 1 ns, and an output range of 0 to -5 V. The amplified signal from the preamplifiers is digitized by a four-channel digital phosphor oscilloscope-computer, Tektronix, TDS 7104, (see Figure 13.6). The incorporated computer finally reads the digitized signal. Figure 13.5 shows the

spectrometer with the fiber, computer, and oscilloscope.

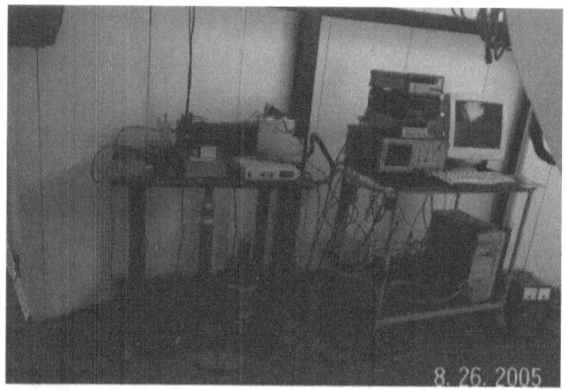

Figure 13.5. The data acquisition electronics.

Figure 13.6. The high resolution Tektronix 7104 oscilloscope
- computer.

13.1.3.4. The filter

A filter can be used instead of the spectrometer to find the maximum signal. The filter, as seen in Figure 13.7, is composed of a coaxial cylindrical adjustable tube that contains a collimator lens, a filter holder, and a single photomultiplier tube designed to maximize the return signal.

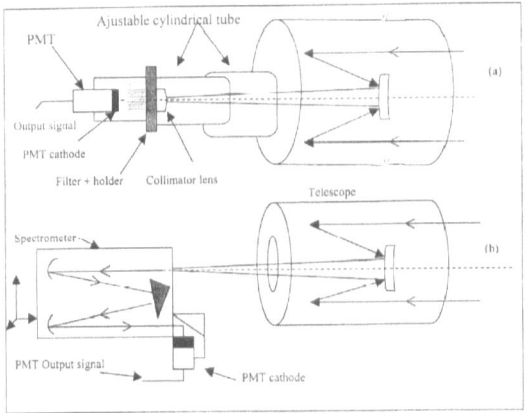

Figure 13.7. (a) The filter system, including the PMT Hamamastu R759, a single interference filter, a collimator lens, and an adjustable telescopic tube, attached to the back of the telescope; (b) layout of

the spectrometer capturing the backscatter from the back of the telescope.

13.2. Experimental procedure
13.2.1. The block diagram of the lidar system

The block diagram is composed of the laser device as the transmitter; the telescope as the receiver system pointing into the atmosphere; the data collection system composed of the oscilloscope, the PMT, the high-voltage power supply, the photodiode, and the power preamps (see Figure 13.8).

13.2.2. The SOLEX method

In this section we will talk about the powerful SOLEX method used with the 248 nm KrF laser system. Figures 13.8 and 13.9 illustrate the pictorial diagram of the SOLEX method. The laser power is continuously monitored during the data acquisition. As in Figure 13.8, a photodiode near the mirror picks a small portion of the laser light and then directs it to the oscilloscope, so as to trigger the data acquisition system. Simultaneously, the mirror reflects the light pulses generated by the KrF excimer laser into the telescope FOV in the atmosphere. As the laser light propagates in the atmosphere, it interacts with molecules and particulates. Then, the scattering process occurs and light is scattered in all directions. The telescope collects the backscatter radiation caused by Rayleigh, Mie, and Raman scattering. In one embodiment, the spectrometer separates the collected signal and directs the wavelength of interest to the PMT. Preamplifiers connected to the output of the PMT amplify the signal from the spectrometer system. The pre-triggered oscilloscope accumulates the amplified signal and records thousands of laser shots that are sent into the computer storage system.

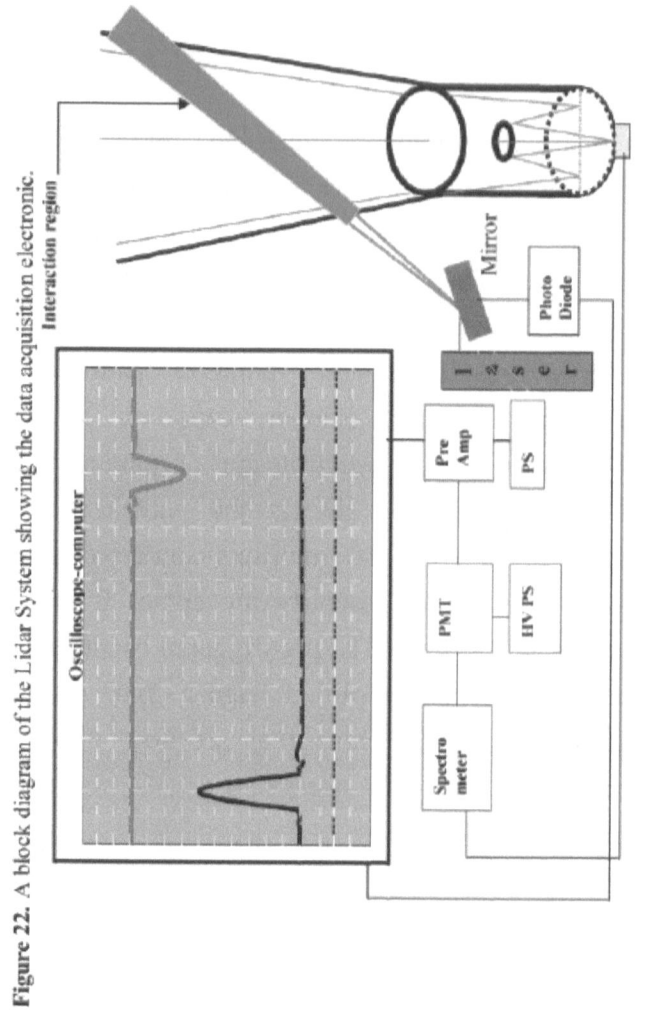

Figure 13.8. Block diagram of the lidar system showing the data acquisition system

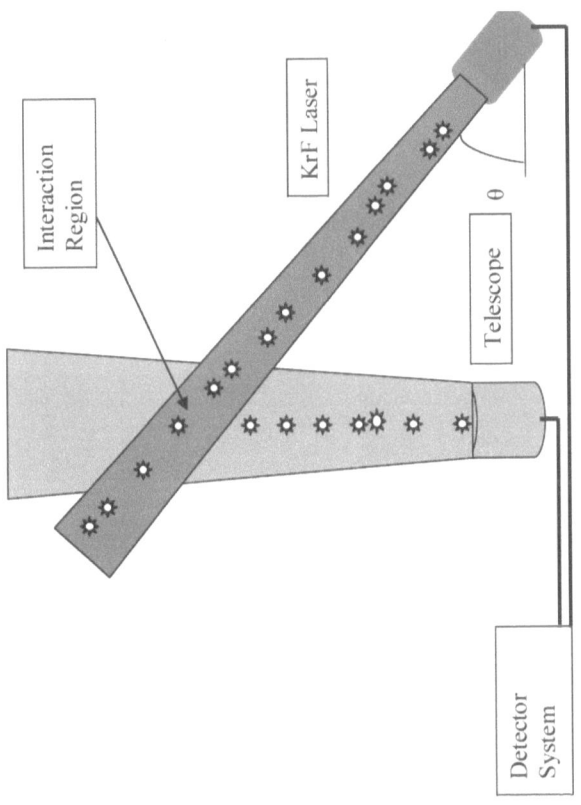

Figure 13.9. A pictorial diagram for the SOLEX method

Chapter 14

Calibration and introduction to data analysis

14.1. Introduction to calibration and data analysis

14.2. Data signal from high resolution Tektronix 7104 oscilloscope.

14.3. The absorption spectrum of water and water vapor

14.1. Introduction to calibration and data analysis

In order to calibrate a spectrometer, the mercury light can be used. Some experimental values of the wavelength of the maximum peaks and their corresponding true value from scientific literature[45] are given in Table 14.1.

Standard Wavelength λ_o (nm)	Experimental Wavelength λ_e (nm)	$\delta\lambda = \lambda_0 - \lambda_e$ (nm)	$\%\lambda$
253.65	253.711	0.061	0.024
265.369	265.364	0.005	0.0019
296.728	296.592	0.136	0.0458
365.01	365	0.01	0.0027
404.656	404.942	-0.286	0.0706
435.833	435.862	-0.029	0.0066
546.074	546.014	0.06	0.0109

Table 14.1. Comparison of the standard wavelength (λo) and the experimental wavelengths (λe)
(Data taken by the author himself with a Beckman spectrometer)

$\delta\lambda$ is the individual deviation.

For a true value of the wavelength λo and the experimental values from the spectrometer, λe, we get the individual deviation δ λ and the experimental error by

$$\%\lambda = \frac{|\delta\lambda|}{\lambda_0} \times 100 \tag{14.1}$$

As can be seen, the experimental deviation in each case is very small, much less than 0.04%. This proves the reliability of the values obtained, and the efficiency of the spectrometer. Figure 14.1 shows the spectrum of the mercury lines, and the spectrometer is shown in Figure 14.2 calibration curve. It is a straight line of slope m = 1 and the y-intercept less than 1 nm. The small maximum deviation on the calibration curve, 0.517 nm and the high correlation coefficient of r = 1, attests the high performance of the spectrometer.

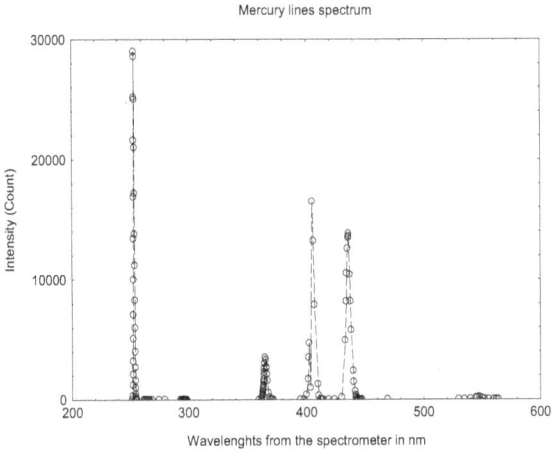

Figure 14.1. Mercury spectrum from the spectrometer

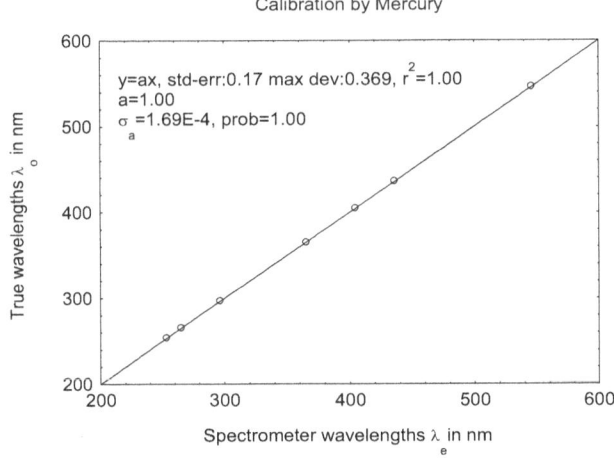

Figure 14.2. Calibration curve of the spectrometer

Figure 14.3 shows the backscatter signal intensity in mV as a function of the wavelength minus 250 nm.

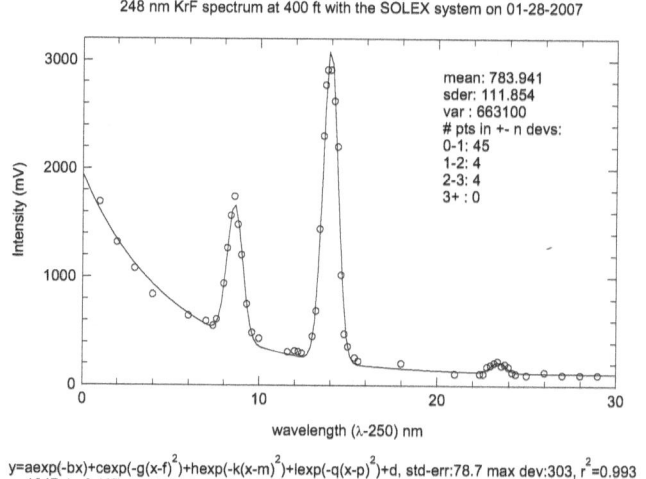

$y=a\exp(-bx)+c\exp(-g(x-f)^2)+h\exp(-k(x-m)^2)+l\exp(-q(x-p)^2)+d$, std-err:78.7 max dev:303, r^2=0.993
a=1847, b=0.197, c=1267, d= 96.7, f=8.60, g=2.72, h=2895, k=2.77, l=95.1, m=14.0, p=23.4, q=2.67
σ_a =88.5, σ_b =0.0184, σ_c =55.0, σ_d =29.5, σ_f =0.0219, σ_g =0.337, σ_h =51.0, σ_k =0.123

Figure 14.3. Raman backscatter for oxygen, Nitrogen and water vapor at 400 ft on the day 01-28-2007 at Beltsville, MD. On the wavelength axis, 250 nm is subtracted from the wavelength.

These data were taken by the author at Beltsville, MD. There are used here to illustrate data analysis. A software program EasyPlot[TM] [56] (www.spiralsw.com/EasyPlot) helped to fit the spectrum with a Multiple Gaussian functions combined with a simple decaying exponential function and a constant d representing the background, on the form:

$$y = a\exp[-bx] + c\exp[-g(x-f)^2] \\ + h\exp[-k(x-m)^2] + l\exp[-q(x-p)^2] + d \quad (14.2)$$

where $a, b, c, d, e, f, g, h, k, l, m, p, q$ are constant real numbers.

EasyPlot[TM] also gave all the coefficients involved in the expression of that function. For the previous case, $a = 1847$ mV, $b = 0.197$ nm^{-1}, $c = 1267$ mV, $d = 96.7$ mV, $f = 8.60$ nm, $g = 2.72$ nm^{-2}, $h = 2895$, $k = 2.77$ nm^{-2}, $l = 95.1$ mV, $m = 14.0$ nm, $p = 23.4$ nm, $q = 2.67$ nm^{-2}.

EasyPlot[TM] gave the maximum deviation and the correlation coefficient r.

In the spectrum obtained on the night of 01-28-2007, these values were maximum deviation: 303 mV, $r^2 = 0.993$. Figure 14.3 proves the ability of our LIDAR system to measure Raman peaks at 250.6 nm for oxygen, 263.9 nm for Nitrogen, and 273.5 nm for water vapor. The first exponential term represents the decaying exponential function c, h, and l represent, respectively, the amplitude of the oxygen peak, the nitrogen peak, and water vapor peak; f, m, and p represent the center wavelength respectively of the first, second, and third Gaussian peaks. The ratio h / c represents the ratio of nitrogen signal with respect to oxygen signal, whereas, l / h represents the ratio of water vapor signal with respect to nitrogen signal. The Raleigh peak is obtained and fitted by a simple Gaussian function from the same software, EasyPlot^{TM} (See Figure 14.4), with a good correlation coefficient $r = 0.996$

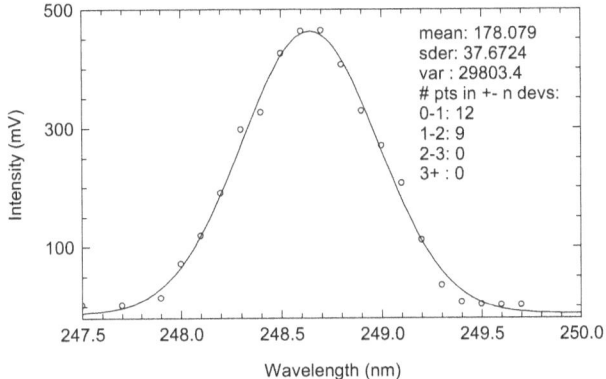

Figure 14.5. Background of the 248 nm KrF Raman lidar

Figure 14.6 illustrates the data points obtained on 01-09-2006. As it is, at the time of 11:35 PM, the atmosphere was not stable. There was some inversion in the lower atmosphere. Figure 30 shows a stable atmosphere on 01-19-1006. Among some other factors that can affect data, are the particulates. Particulate matter, also known as particle pollution or PM is a complex mixture of extremely small particles and liquid droplets. Particle pollution is made up of a number of components including acids (such as nitrates and sulfates), organic chemicals metals, and soil or dust particles. According to the US Environmental Protection Agency (EPA), even particulate matter of the order of a microgram per cubic-meter can be dangerous for health.

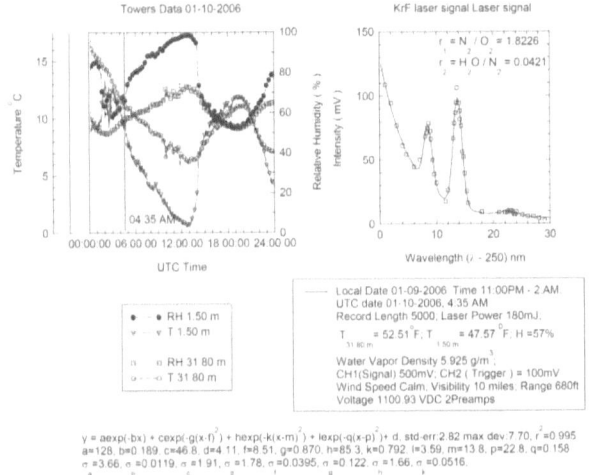

Figure 14.6. Beltsville, MD 248 nm KrF LIDAR data and tower data plots. Date: 01-09-2006.

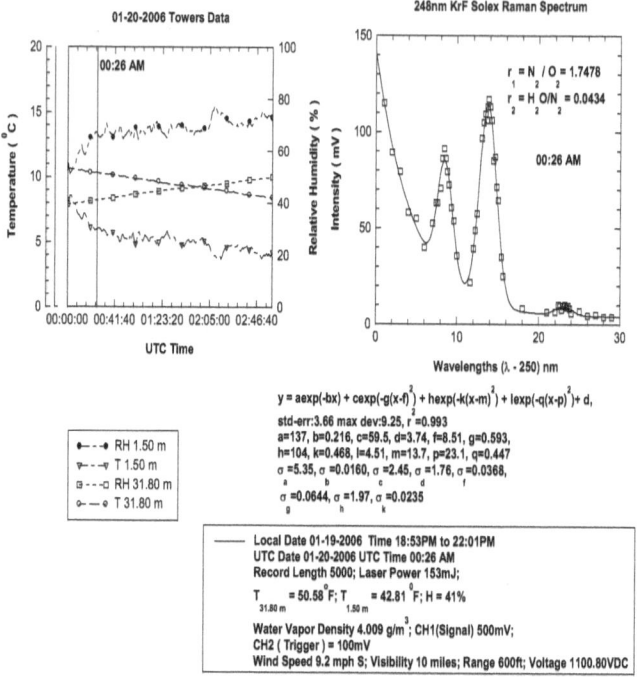

Figure 14.7. Beltsville 248 KrF LIDAR data and tower data plots Date: 01-19-2006.

14.2. Data signal from high resolution Tektronix 7104 oscilloscope.

In the following pictures are some oscilloscope's signals obtained from water vapor, nitrogen and oxygen.

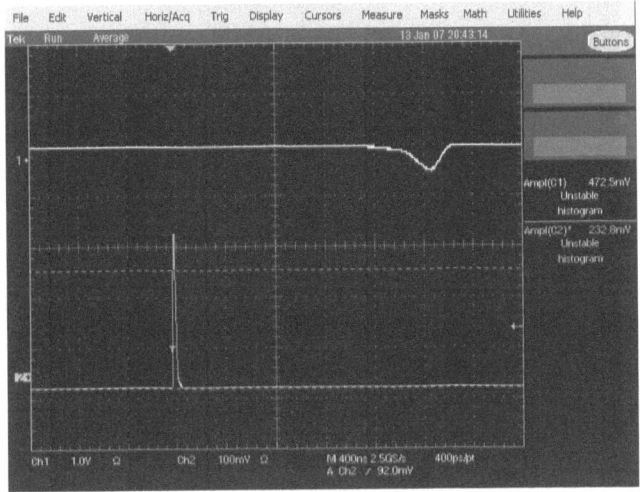

Figure 14.8. Raleigh signal at 248.6 nm on 01-13-2007 at 20:43:09.

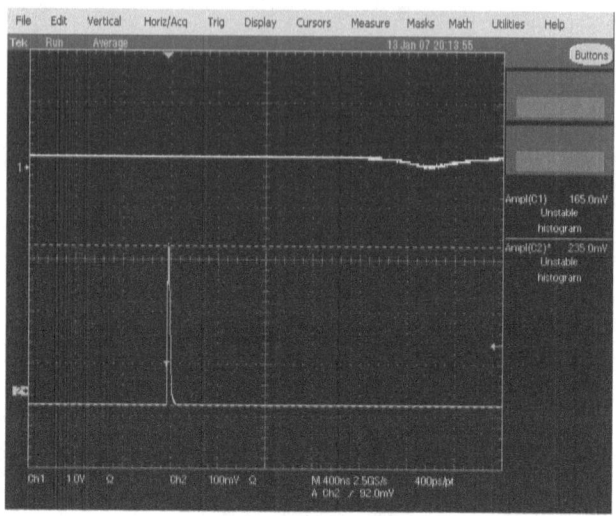

Figure 14.9. Oxygen signal on 01-13-2007 at 20:13:55.

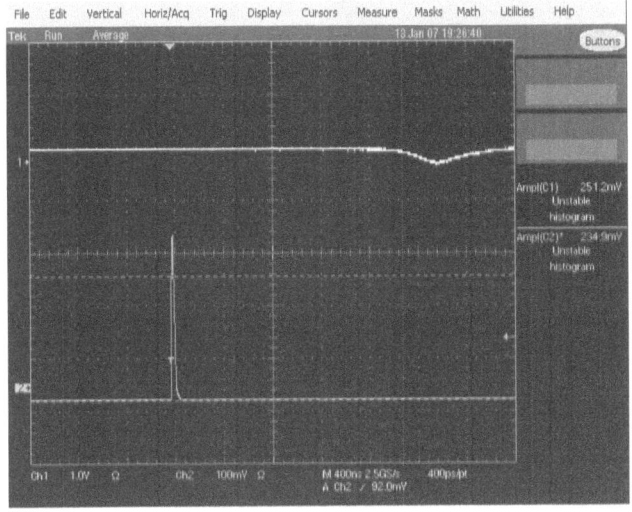

Figure 14.10. Nitrogen signal on 01-13-2007 at 19:26:40.

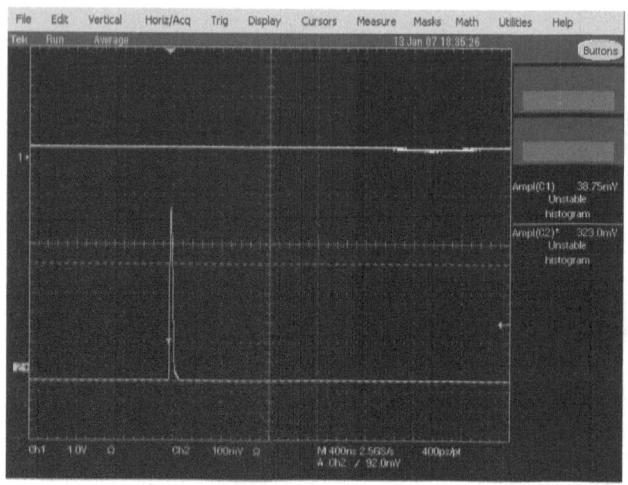

Figure 14.11. Water vapor signal on 01-13-2007 at 18:35:26.

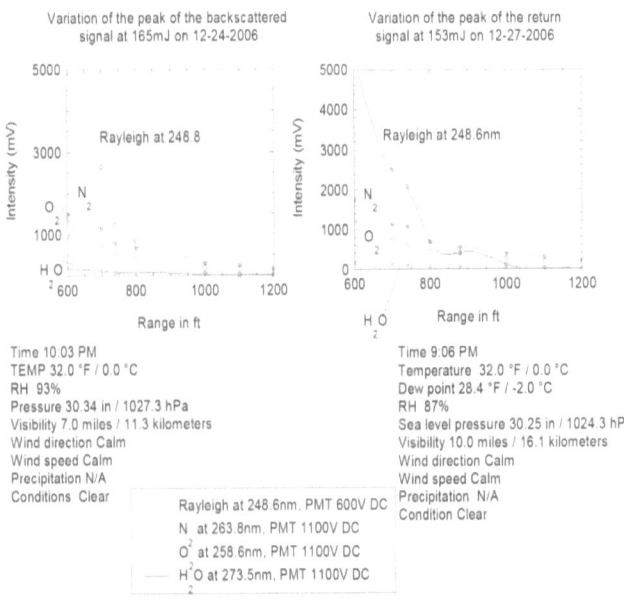

Figure 14.12. Variation of the power supply intensity as a function of the range from 600 ft to 1200 ft.

14.3. The absorption spectrum of water and water vapor

The atmospheric absorption spectrum is complicated. Indeed there are several particles and chemical reactions in the atmosphere. On one hand, nitrogen and oxygen do not contribute too much to the absorption radiation. On the other hand, variable constituents of the atmosphere such as water vapor, H_2O, carbon dioxide, CO_2, Ozone, O_3, hydrocarbon combinations and nitrogen oxides have absorption lines and bands in some given spectral regions. Figure 14.13 shows the spectral absorption lines of some atmospheric gases from the UV to the microwave region. The absorption line of the water vapor is so large that many investigations have been done on the structure of water vapor molecules and the origins of these absorption lines.

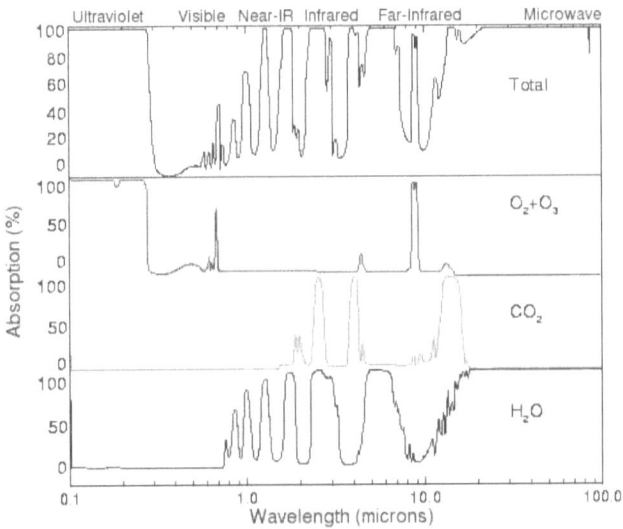

Figure 14.13. Absorption lines for water vapor, carbon dioxide, oxygen and ozone

Investigations have proved that in the unexcited state, water vapor molecule configuration has the form of an isosceles triangle as shown in figure 14.14 and 14.15

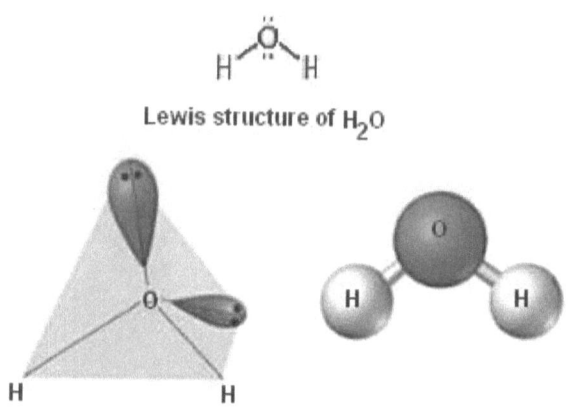

Figure 14.14. Water vapor molecule structure

In the spectral region $[0.54\mu m; 9\mu m]$ the absorption of water vapor is caused by vibrational rotational transitions. However, in the spectral regions $[9\mu m; 1.5 cm]$, or the far infrared it is caused by purely rotational transitions. In the Ultraviolet, water vapor absorption is caused by electronic transitions. The frequencies for the three types of normal vibrations for water vapor are:

$v_1 = 3670 cm^{-1}$; $v_2 = 1675 cm^{-1}$; $v_3 = 3790 cm^{-1}$.

The rotational absorption spectrum is due to the rotational energy transitions of water vapor molecules with respect to the three axis of rotation asymmetrically. We will now observe the situation where water vapor absorption is done by shortwave (solar) radiation

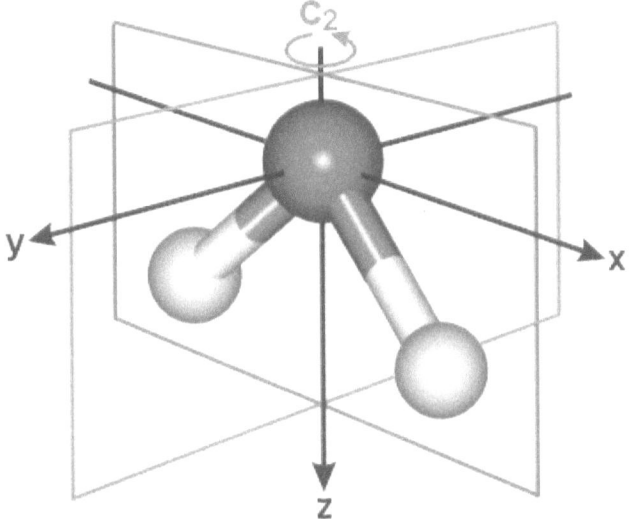

Figure 14.15. Water vapor molecules structure and rotational axis

14.4. Shortwave (solar) radiation absorption by water vapor

Water vapor has a number of intensive absorption bands in the far ultraviolet. Some of these intensive bands can be seen in the following regions:

$160 \overset{o}{A} < \lambda < 1100 \overset{o}{A}$, $\quad 1050 \overset{o}{A} < \lambda < 1450 \overset{o}{A}$,

$1450 \overset{o}{A} < \lambda < 1900 \overset{o}{A}$. The absorption of water vapor by the UV solar radiation is a major contribution to energy in the upper atmosphere. Meanwhile, by the far UV radiation, the absorption of water vapor appears to be insignificant in the troposphere - these radiations being almost completely absorbed in the upper atmosphere. In the visible, especially in the wavelength region $572nm < \lambda < 703nm$, there is also a very small absorption lines by water vapor known in this case as rain. In the near infrared, there are a number of intensive absorption bands by the water vapor. The following table shows the different intervals in which the maximum absorption occurs.

Spectral regions in nm	Band Center Maximum Absorption	Band's denomination
700-740	718	α
790-840	810	0.8μ
926-978	935	ρ, σ, τ
1095-1165	1130	φ
1319-1498	1395	ψ
1762-1977	1870	Ω
2520-2845	2680	χ

Table 14.2 Spectral regions of intensive water vapor absorption in the near and middle infrared radiation

14.5. Computation

Scientists such as Yamamoto and Onishi[72,73] has computed the absorption by water vapor in longer wavelength such as near 6,300 nm and 3200 μm where there's intensive absorption. The absorption radiation depends on the pressure of the medium and its water vapor content. From Howard et al.[74,75], the following formula can be used:

For weak absorption:

$$\int_{v_1}^{v_2} A_v dv = cw^{1/2}(p+e)^k \qquad \int_{v_1}^{v_2} A_v dv < A_c \quad (14.3)$$

For strong absorption:

$$\int_{v_1}^{v_2} A_v dv = C + D\log w + K\log(p+e)$$

$$\int_{v_1}^{v_2} A_v dv > A_c \qquad (14.4)$$

where p is the total pressure in mm, e is the partial pressure of water vapor in mm, w is the water vapor content on the ray's path, v_1 and v_2 are the absorption band lower and upper limits. c, C, D, k,

K are empirical constants. The mean absorption function $\overline{A}$ is given by the relation:

$$\overline{A} = \frac{1}{\Delta v}\int A_v dv \qquad (14.5)$$

From data and analyses, Möller[76] and Mügge were able to derive an empirical formula to calculate solar radiation flux absorbed in a cloudless atmosphere in terms of the total water vapor content in the atmosphere:

$$\Delta S = 0.172(mw_\infty)^{0.3}$$

where m is the atmospheric mass directed to the sun, w_∞ the total water vapor content in the atmosphere in the vertical direction in g/cm^2.

A more precise formula was later derived by Möller et al.[77] based on radiation data by Howard et al. as follows:

$$\Delta S = \exp\{2.3026[-0.740 + 0.347\log(mw_\infty) \\ -0.056\log(mw_\infty)^2 - 0.006\log(mw_\infty)^3]\} \qquad (14.6)$$

Angström[78] first suggested the formula:

$$\Delta S = 0.10 + 0.21(1 - e^{-0.23mw_\infty}) \qquad (14.7)$$

But later assume that all these formula which take into account only pure absorption cannot be applied for the calculation of the solar radiation absorption in the real atmosphere[79]. The scattering of the solar radiation has also to be taken into account. Therefore the following formula was used:

$$\Delta S = 0.10 + \frac{0.23 w_\infty}{0.23 w_\infty + \beta} 0.21(1 - e^{-m\beta} e^{-0.23 m w_\infty}) \quad (14.8)$$

β is the optical atmospheric mass caused by scattering.

Appendices

Appendix A: Saturated Vapor pressure variation as a function of the temperature in °F.

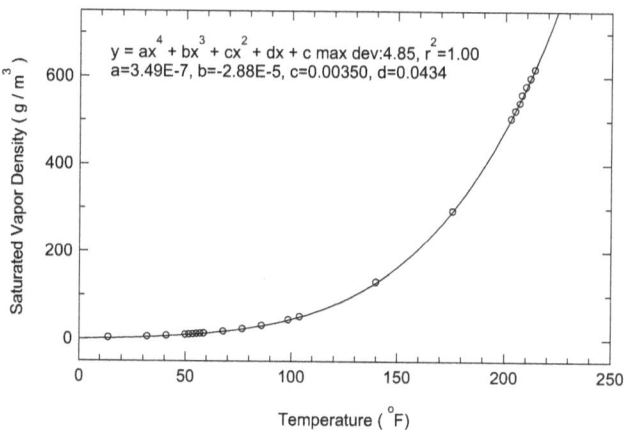

Figure A. Saturated vapor density vs. temperature.

Appendix B: LIDAR data and spectra

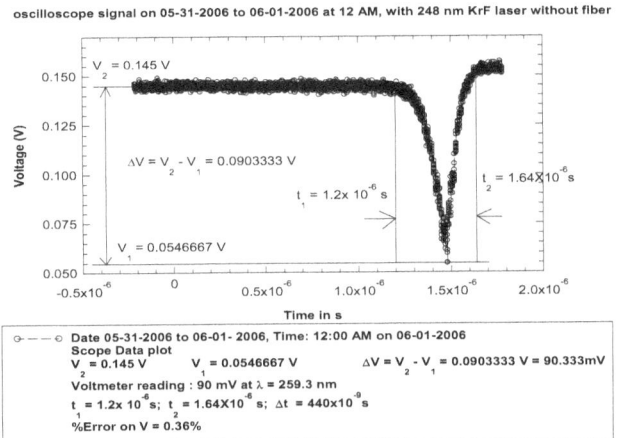

Figure B$_1$. The example of the 248 KrF laser return signal from the oscilloscope.

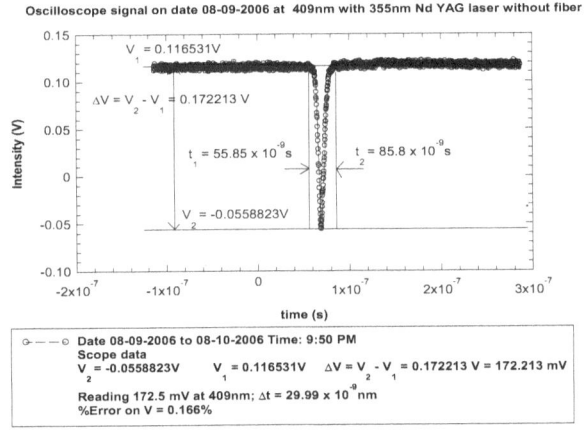

Figure B$_2$. The 355 Nd:YAG laser return signal from the oscilloscope

List of symbols, abbreviations, initialisms and acronyms

Lidar: light detection and ranging
Laser: light amplification by the stimulated emission of radiation
PMT: photomultiplier tube
SI: System International Unit
SOLEX: Selective Overlap Lidar Experiment
^oT: Temperature
^oF: Degree Fahrenheit
^oC: Degree Celsius
RH: relative humidity
Wind dir. wind direction
MD: Maryland
FOV: field of view
ρ_v: water vapor density
XeF: xenon fluoride
KrF: krypton fluoride
Nd:YAG: neodymium yttrium aluminum garnet
ppb: parts per billion
EPA: U.S. Environmental Protection Agency
G: ground level

References

1. David Whiteman, "Examination of the Traditional Raman LIDAR Technique—I—Evaluating the Temperature-Dependent LIDAR Equations," Applied Optics 42, no. 15 (2003): 2571–2592.

2. "Examination of the Traditional Raman LIDAR Technique—II—Evaluating the Ratios for Water Vapor and Aerosols," Applied Optics 42, no. 15 (2003): 2593–2608.

3. D. Renault, R. Capitini, "Boundary Layer Water-Vapor Probing with a Solar-Blind Raman LIDAR: Validations Meteorological Observations and Prospects," J. Atmos. Oceanic Tech. 5 (1988): 5-15.

4. Kamil Stelmaszczyk, Marcella Dell'Aglio, Stanislaw Chudzynski, Tadeusz Stacewicz, and Ludger Wöste, "Analytical Function for LIDAR Geometrical Compression Form-Factor Calculations," Applied Optics 44, no. 7 (2005): 1323.

5. R. Ferrare, D. D. Turner, M. Clayton, S. Guilbert, M. Schulz, and M. Chin, "The Vertical Distribution of Aerosols over the Atmospheric Radiation: Measurement Southern Great Plains Site Measured versus Modeled" (Fifteenth ARM Science Team

Meeting Proceedings, Daytona Beach, FL, March 14–18, 2005).

6. Andreas Behrendt, Gerd Wagner, Anna Petrova, Max Shiler, Sandip Pal, Thorsten Schaberl, and Volker Wulfmeyer, "Modular LIDAR Systems for High-Resolution 4-Dimensional Measurements of Water Vapor, Temperature, and Aerosols," Institute of Physics and Meteorology, University of Hohenheim, Stuttgart, Germany, 2005.

7. Brian J. Soden, David D. Turner, Barry M. Lesht, and Larry M. Miloshevich, "An Analysis of Satellite, Radiosonde, and LIDAR Observations of Upper Tropospheric Water Vapor from the Atmospheric Radiation Measurement Program," Journal of Geophysical Research 109, D04105, February 15, 2004.

8. Olivier Bock, Jerome Tarniewicz, Yann Morille, Jacques Pelon, and Christian Thom, "Developpement d'un LIDAR Raman pour le sondage de la vapeur d'eau et la correction des delais tropospheriques en GPS," Bulletin d'Information de l'IGN 74 (2003): 3.

9. Ahmed M. Farah, Demetrius D. Venable, Arthur N. Thorpe, Frederick Marsh, and William S. Heaps, "Validation of a Novel Ultraviolet LIDAR System

with Relative Raman-Scattering Cross Sections Determined from Atmospheric Measurements," Applied Optics 41, no. 3 (2002): 407–411.

10. Ina Mattis, Albert Ansmann, Dietrich Althausen, Volker Jaenisch, Ulla Wandinger, Detlef Müller, Yuri F. Arshinov, Sergej M. Bobrovnikov, and Ilya B. Serikov, "Relative-Humidity Profiling in the Troposphere with a Raman LIDAR," Applied Optics 41, no. 30 (2002): 6451–6462.

11. Geary K. Schwemmer, David O. Miller, Thomas D. Wilkerson, and Ionio Andrus, "Time Resolved 3-D Mapping of Atmospheric Aerosols and Cloud During the Recent Atmospheric Radiation Measurement Water Vapor Intensive Operating Period," LIDAR Remote Sensing for Industry and Environmental Monitoring II, Proceedings of SPIE, vol. 4484, SPIE, (2002): 0277-786X.

12. John R. Ferrero and Kazuo Nakamoto, "Introductory Raman Spectroscopy", Harcourt Brace Academic Press (1994).

13. Raymond M. Measures, "Laser Remote Sensing Fundamental and Applications", New York: John Wiley (1984).

14. Didier Bruneau, Philippe Quaglia, Cyrille Flamant, Mireille Meissonnier, and Jacques Pelon, "Airborne LIDAR Leandre II for Water-Vapor Profiling in the Troposphere. I. System Description," Applied Optics 40, no. 21 (2001): 3450–3461.

15. Benoit Lazzarotto, Max Frioud, Gilles Larcheveque, Valentin Mitev, Phillipe Quaglia, Valentin Simeonov, Anne Thompson, Hubert Van den Bergh, and Bertrand Calpini, "Ozone and Water Vapor Measurements by Raman LIDAR in the Planetary Boundary Layer: Error Sources and Field Measurements," Applied Optics 40, no. 18 (2001): 2985-2997.

16. Chiao-Yao She, "Spectral Structure of Laser Light Scattering Revisited: Bandwidths of Nonresonant Scattering LIDARs," Applied Optics 40, no. 27 (2001): 4875-4884

17. C. M. Penny and M. Lapp "Raman scattering cross sections for water vapor," J. Opt., Soc. Am.,Vol.66No 5 (1976): 422–425.

18. D. N. Whiteman, K. D. Evans, B. Demoz, D. O'C. Starr, E. W. Eloranta, D. Tobin, W. Feltz, G. J. Jedlovec, S. I. Gutman, G. K. Schwemmer, M. Cadirola, S. H. Melfi, and F. J. Schmidlin, "Raman LIDAR Measurements of Water Vapor and Cirrus Clouds During the passage of Hurricane Bonnie," Journal of Geophysical Research 106, no. D6 (2001): 5211–5225.

19. Fernando de Tomasi, Maria R. Perrone, and Maria L. Protopapa, "Monitoring O_3 with Solar-Blind Raman LIDARS," Applied Optics 40, no. 9 (2001): 1314-1320

20. David N. Whiteman, Geary Schwemmer, Timothy Berkoff, Henry Plotkin, Luis Ramos-Izquierdo, and Gelsomina Pappalardo, "Performance Modeling of an Airborne Raman Water-Vapor LIDAR," Applied Optics 40, no. 3, (2001): 375-390.

21. F. de Tomasi, G. Torsello and M. R. Perone, "Water-vapor Mixing-Ratio Measurements in the Solar-Blind Region," Optics Letters, Vol. 25, Issue 10, (2000): 686-688.

22. B. Lazzarotto, V. Simeonov, P. Quaglia, G. Larcheveque, H. Van den Bergh, and B. Calpini, "Ozone and Water Vapour by Raman LIDAR in the

PBL," IPSL Publication Twentieth International Laser Radar Conference, July 10–14, 2000. Vichy-France

23. Benoit Lazzarotto, Gilles Larcheveque, Phillipe Quaglia, Valentin Simeonov, Hubert Van den Bergh, and Bertrand Calpini, "Raman DIAL Measurement of Ozone and Water Vapor in the Lower Troposphere," EPFL, Lausanne, Switzerland, 1999.

24. Voker Wulfmeyer and Jens Bösenberg, "Ground-Based Differential Absorption LIDAR for Water-Vapor Profiling: Assessment of Accuracy, Resolution and Meteorological Applications,", Applied Optics 37, no. 18 (1988):

25. J. E. M. Goldsmith, Forest H. Blair, Scott E. Bisson, and David D. Turner, "Turn-key Raman LIDAR for Profiling Atmospheric Water Vapor, Clouds, and Aerosols," Applied Optics 37, no. 21 (1998): 4979–4990.

26. Dukhyeon Kim, Hyungki Cha, Jongmin Lee, Kyuhwang Yeon, Sungchul Choi, "Determination of Water Vapor and Aerosol Densities in the Tropospheric Atmosphere from Nitrogen and Water Vapor Raman Signals" Journal of Korean Physical Society, Vol. 33, No 3 (1998): 301-307

27. Scott E. Bisson, "Parametric Study of an Excimer-Pumped, Nitrogen Raman Shifter for LIDAR Applications," Applied Optics 34, no. 18 (1995): 3406–3412.

28. William E. Eichinger, Daniel I. Cooper, Fred L. Archuletta, Douglas Hof, David B. Holtkamp, Robert R. Karl Jr., Charles R. Quick, and Joseph Tiee, "Development of a Scanning Solar-Blind, Water Raman LIDAR," Applied Optics 33, no. 18 (1994): 0003-6935

29. D. N. Whiteman, S. H. Melfi, and R. A. Ferrare, "Raman LIDAR System for the Measurement of Water Vapor and Aerosols in the Earth's Atmosphere," Applied Optics 31, no. 16 (1992): 3068–3082

30. W. C. Bain and M. C. Sandford, "Light Scatter from a laser Beam at Height Above 40 km," Radio and Space Research Station, Ditton Park Slough, Journal of Atmospheric and Terrestrial Physics, Vol. 28, no. 6-7 (1966): 543-552.

31. Donald A. Leonard, "Observation of Raman Scattering from the Atmosphere Using a Pulsed Nitrogen Ultraviolet Laser," Nature, Vol. 216, Issue 5111, (1967) 142-143

32. Louis Elterman and Allan B. Campbell, "Atmospheric Aerosol Observations with Searchlight Probing," *Journal of Atmospheric Sciences* 21 (July 1964): 457–458.

33. O K Kostko, "Use of laser radar in atmospheric investigations (review)," *Sov. J. Quantum Electron,* Vol. 5(10) (1975): 1161-1177

34. Franz Balsiger, C. Russell Philbrick, "Comparison of Lidar Water Vapor Measurements Using Raman Scatter at 266 nm and 532 nm," *Proc. SPIE,* Vol. 2833, 231 (1996): 1826–1829.

35. John Cooney, Kenneth Petri, and Alfred Salik, "Measurement of High Resolution Atmospheric Water-Vapor Profiles by the Use of a Solar Blind Raman LIDAR," Applied Optics 24, no. 1 (1985): 104–108.

36. D. Renaut, J. C. Pourny, R. Capitini, "Daytime Raman—LIDAR Measurement of Water Vapor," Optics Letters 5, no. 6 (1980): 0146-9592/80/060233-03

37. A. T. Young, "On the Rayleigh-Scattering Optical Depth of the Atmosphere," Journal of Applied Meteorology vol. 20, Issue 3, (1981): 328–330

38. W. J. Wiscombe, "Improve Mie Scattering algorithms" Applied Optics, Vol. 19, Issue 9, (1980): 1505-1509

39. Rupa Kamini, T. N. Krishnamurti, Richard A. Ferrare, Syed Ismail, and Edward V. Browell, "Impact of High Resolution Water Vapor Cross-Sectional Data on Hurricane Forecasting," Geophysical Research Letters 30, no. 5 (2003): 38.1-38.4

40. Scott E. Bisson, John E. M. Goldsmith, and Mark G. Mitchell, "Narrow-Band, Narrow Field-of-View Raman LIDAR with Combined and Night Capability for Tropospheric Water-Vapor Profile Measurements," Applied Optics 38, no. 9 (1999): 1841-1849.

41. J. Harms, "LIDAR Return Signals for Coaxial and Noncoaxial Systems with Central Obstruction," Applied Optics 18, no. 10 (1979): 1559-1566.

42. Robert J. Hall, John A. Shirley, and Alan C. Eckbreth, "Coherent Anti-Stokes Raman Spectroscopy: Spectra of Water Vapor in Flames," Optics Letters, Vol. 4, Issue 3, (1979) 87-89

43. S. H. Melfi, J. D. Lawrence Jr., and M. P. Mccormick, "Observation of Raman Scattering by

the Water Vapor in the Atmosphere," Applied Physics Letters, Vol. 15 (1969): 295-297.

44. J. A. Cooney, "Measurements on the Raman Component of Laser Atmospheric Backscatter," Radio Corporation of America, Astro-Electronics Division Princeton, NJ, Applied Physics Letters, Vol. 12 (1968): 40-42.

45. Robert C. Weast and Melvin Astle, "Handbook of Chemistry and Physics," 62nd ed., E-260 / E261, (1982).

46. Gerhard Herzberg, "Molecular Spectra and Molecular Structure—Spectra of Diatomic Molecules," 2nd ed. F.R.S. National Research Council of Canada. Book, (1950).

47. Hassan Moore, "Sensitivity Verification for a Non-coaxial LIDAR Configuration for the Analysis of Ozone", PhD dissertation, Howard University (2006).

48. Vladimir A. Kovalev and William E. Eichinger, "Elastic LIDAR-Theory, Practice, and Analysis Methods" New Jersey: Wiley Interscience, John Wiley (2004).

49. Takashi and Tetsuo Fukuchi, "Laser Remote Sensing," (CRC Press, Taylor & Francis Group,

6000 Broken Sound Parkway NW, Suite 300 Boca Raton, FL (2005).

50. Orazio Svelto and David C. Hanna, "Principles of Lasers" 2nd ed. New York: Plenum Press, (1982).

51. Ahmed M. Farah, "Development of an Excimer Laser-based Raman LIDAR For tropospheric Ozone Concentration Measurements" (PhD diss., Howard University, 2001).

52. Upendra N. Singh and Kohei Mizutani, "LIDAR Remote Sensing for Industry and Environmental Monitoring V," Proceedings of the SPIE 4153, Vol. 5653 (2005)

53. J. A. Cooney, "Remote Measurements of Atmospheric Water Vapor Profiles Using the Raman Component of Laser Backscatter," J. Appl. Meteorol. 9 (1970): 182–183.

54. William T. Silfvast, Laser Fundamentals, Publish in the United States of America, by Cambridge University Press, New York (1996).

55. R. W. Engstrom, Photomultiplier Handbook: Theory, Design, Application, Burle Technology, RCA, Princeton, N.J., (1980)

56. Stuart Karon, "EasyPlot for Microsoft Windows Scientific Plotting and Data Analysis," Spiral

Software and Massachusetts Institute of Technology (1988–1997).

57. G. Fiocco and L. D. Smullin, "Detection of Scattering Layers in the Upper Atmosphere (60–140 km) by Optical Radar," Nature 199, (1963): 1275 – 1276.

58. John R. Reitz, Frederick J. Milford, Robert W. Christy, "Foundations of Electromagnetic Theory", 3rd ed. Addison-Wesley, New Jersey (1979).

59. Daniel R. Frankl, "Electromagnetic Theory" Englewood Cliffs, NJ: Prentice Hall, (1986).

60. William M Steen, "Laser Materials Processing," 2nd ed., Springer-Verlag, New York, (1998).

61. Ivar Waller, Presentation Speech. "The Nobel Prize in Physics 1966", Nobel lectures Physics 1963-1970, Elsevier Publishing Company, Amsterdam, (1972).

62. R. Gordon Gould, "The LASER, Light Amplification by Stimulated Emission of Radiation," in the Ann Arbor Conference on Optical Pumping, University of Michigan, (1959).

63. J.T. Houghton, G.J. Jenkins, J.J. Ephraums, "In Climate Change: The IPCC Scientific Assessment. WMO/UNEP Report of Working group I of the Intergovernmental Panel on Climate Change,"

(Eds.). Cambridge University Press, Cambridge, UK, (1990).

64. Albert Ansmann, Maren Riebesell, Claus Weitkamp, "Measurement of atmospheric aerosol extinction profiles with a Raman lidar," Opt. Lett. 15 (1990): 746-748.

65. Syukuro Manabe and Richard T. Wetherald, "Thermal Equilibrium of the Atmosphere with a Given Distribution of Relative Humidity" Journal of Atmospheric sciences, Vol.24, No 3 (1967): 241-259

66. H. W. Ellsaesser, "The Climatic Effect of CO_2: A Different View." Atmos. Env., 18, (1984): 431-434.

67. W.G. Rees. Physical Principles of Remote Sensing. Cambridge University Press; 1990.

68. Sears and Zemansky's University Physics with Modern Physics, Young and freedman, 13th Edition, Pearson education, Inc. 2012, 2008, 2004

69. K.D. Moller. Optics. Learning by Computing with examples Using Mathcad, 2003 Springer-Verlag New York, Inc.

70. pollo.lsc.vsc.edu/classes/met130/notes/chapter1/vert_temp_all.html

71. Richard Cadle. Particles in the atmosphere and space. Reinhold publishing corporation, New York, 1966.

72. G. Yamamoto, G. Onishi, Absorption coefficient of water vapor in the far infrared region. Sci. Rept. Tohoku Univ., Fifth Ser. 1, No 1; 1949.

73. G. Yamamoto, G. Onishi. Absorption coefficient of water vapor in the near infrared region. Sci. Rept. Tohoku Univ., Fifth Ser. 1, No 1; 1949.

74. J.N. Howard, D. E. Burch, D. Williams. Near infrared transmission through synthetic atmospheres. J. Opt. Soc. Am. Nos. 3,4 and 5. 1956.

75. J.N. Howard, D. E. Burch, D. Williams. Near-infrared transmission through synthetic atmospheres. Ohio State Univ., Geophys. Res. Paper No. 40. 1955.

76. F. Möller. Strahlung in der unteren Atmosphäre. Springer, Berlin, 1957.

77. F. Kasten, G. Korb, G. Manier, F. Möller, On the heat balance of the troposphere, Final Rept., Contrat AF 61(052)-18. Mainz, 1959.

78. A. Angström, Absorption of solar radiation by atmospheric water vapor. Arkiv Geofysik 3, No. 23, 1961.

79. K.YA. Kondratyev, Radiation in the atmosphere, Academic Press, New York and London, 1969

INDEX

A
Amplifier 180,191,193,204
Atmosphere 26-35
Atmospheric 232

B

C
Calibration 260,262
Coulomb 51, 300
-, law 51

D
Differential operator 62
Dynode 243,245-247

E
Earth 2, 16-20
Electromagnetic 40,243
-energy 53,54,56,57
-wave equations 57
Electromagnetism 60

-, Basic laws of
Emission 9,143, 177, 178, 180, 182, 182, 146,-284-286
, spontaneous 178, 183
-, stimulated 178, 180
Environmental 266,284,286,287,295
Electromagnetic 40,47,48,53,54,56,57,62,63,65,77
Excited state 211,275
Excimer 177,182,210-213
Excimer laser 177,182,210-213
Extinction coefficient 224,226

F
Faraday 48-51

G
Gaussian 263,264

H
Humidity 34,231
-, absolute 10,220,230
-, specific10, 220,229
-, relative 220,229,284
Hurricane289, 293

I
Interferometer, Michelson- 118,129,130
Interferometry 180,210
Inelastic scattering 215-218
Inversion 177, 190-198

K
Krypton fluoride 284

L
Laplace equation 47, 62
Laser
-, Helium Neon 207
-, Neodymium Yag 203
-, Ruby 199-202
-, excimer 210
Law
Ampere's circuital- 48-51, 244
Coulomb's-, 51
Energy conservation- 47, 56, 57
Faraday's- 48-51
Gauss'-, 48-50, 60-62
Oersted's-, 60, 61
Planck's-, 144-146,148,178, 183

Lidar 10, 11, 38,220, 221, 222, 224, 226, 232, 233, 237, 238,241, 254, 256, 264, 268

M
Maser 178-180, 193, 199
Maxwell equations 48-50, 58
Metastable 210
Mie Scattering 215
Mixing 32, 220, 224,228,229
- Ratio 224,228,229

N
Neodymium 177,
-, Yag laser, 204-207, 232, 233, 283
Nitrogen 32,38,207, 224-227, 232,262,264, 268,274,290,291

O
Oxygen 32,38,207, 232
Ozone 43,234,236,274, 275,288,289,290,294

P
Particulate 177, 215, 238, 255, 266
Planetary 19, 288
Population inversion 193-196

Poisson equation 47, 62

Polarization 72-78

R

Radiation 32,40-42 255, 277-281

Raman 177, 211, 215-219, 232, 233, 255

Rayleigh 145, 215-217, 255

Relative Humidity 34, 220, 229, 284

Remote sensing 38-42 232, 287

S

Shortwave radiation 63, 277

Solex 237, 251, 254

Stimulated emission 177-188

Stoke 217, 218

Acquisition system 254, 256

Climate- 234

Collection- 254

Detector- 242

Four-Level- 199

Filter- 242, 246

Receiver- 242, 254

Spectrometer- 255

Transmitter- 238

Telescope- 131, 221, 222, 225, 226, 238

Scattering 213
-, Mie 215
-, Rayleigh 213
-, Raman 215-219

T
Temperature 29-32, 38

U
Ultra-violet 193

V
Vibration 82,83
Vibrational 211, 216-218, 233, 276
Vapor Pressure 24, 220, 229, 231

W
Water vapor
-density 52, 54, 55, 230, 282
-Measurement 20, 129, 232, 286, 287
Wavelength 41, 63, 95, 120, 172, 173, 217, 246
Wind 32, 284

X

Xenon Fluoride 182, 284
Y-Yag 203, 204, 233

www.ingramcontent.com/pod-product-compliance
Lightning Source LLC
Chambersburg PA
CBHW020729180526
45163CB00001B/170